Models in Environmental Regulatory Decision Making

Committee on Models in the Regulatory Decision Process

Board on Environmental Studies and Toxicology

Division on Earth and Life Studies

NATIONAL RESEARCH COUNCIL
OF THE NATIONAL ACADEMIES

THE NATIONAL ACADEMIES PRESS
Washington, D.C.
www.nap.edu

THE NATIONAL ACADEMIES PRESS 500 Fifth Street, NW Washington, DC 20001

NOTICE: The project that is the subject of this report was approved by the Governing Board of the National Research Council, whose members are drawn from the councils of the National Academy of Sciences, the National Academy of Engineering, and the Institute of Medicine. The members of the committee responsible for the report were chosen for their special competences and with regard for appropriate balance.

This project was supported by Contract No. 68-C-03-081 between the National Academy of Sciences and the U.S. Environmental Protection Agency. Additional support was provided by the U.S. Department of Transportation. Any opinions, findings, conclusions, or recommendations expressed in this publication are those of the author(s) and do not necessarily reflect the view of the organizations or agencies that provided support for this project.

International Standard Book Number-13: 978-0-309-11000-6
International Standard Book Number-10: 0-309-11000-9

Additional copies of this report are available from

The National Academies Press
500 Fifth Street, NW
Box 285
Washington, DC 20055

800-624-6242
202-334-3313 (in the Washington metropolitan area)
http://www.nap.edu

Printed in the United States of America.

THE NATIONAL ACADEMIES

Advisers to the Nation on Science, Engineering, and Medicine

The **National Academy of Sciences** is a private, nonprofit, self-perpetuating society of distinguished scholars engaged in scientific and engineering research, dedicated to the furtherance of science and technology and to their use for the general welfare. Upon the authority of the charter granted to it by the Congress in 1863, the Academy has a mandate that requires it to advise the federal government on scientific and technical matters. Dr. Ralph J. Cicerone is president of the National Academy of Sciences.

The **National Academy of Engineering** was established in 1964, under the charter of the National Academy of Sciences, as a parallel organization of outstanding engineers. It is autonomous in its administration and in the selection of its members, sharing with the National Academy of Sciences the responsibility for advising the federal government. The National Academy of Engineering also sponsors engineering programs aimed at meeting national needs, encourages education and research, and recognizes the superior achievements of engineers. Dr. Charles M. Vest is president of the National Academy of Engineering.

The **Institute of Medicine** was established in 1970 by the National Academy of Sciences to secure the services of eminent members of appropriate professions in the examination of policy matters pertaining to the health of the public. The Institute acts under the responsibility given to the National Academy of Sciences by its congressional charter to be an adviser to the federal government and, upon its own initiative, to identify issues of medical care, research, and education. Dr. Harvey V. Fineberg is president of the Institute of Medicine.

The **National Research Council** was organized by the National Academy of Sciences in 1916 to associate the broad community of science and technology with the Academy's purposes of furthering knowledge and advising the federal government. Functioning in accordance with general policies determined by the Academy, the Council has become the principal operating agency of both the National Academy of Sciences and the National Academy of Engineering in providing services to the government, the public, and the scientific and engineering communities. The Council is administered jointly by both Academies and the Institute of Medicine. Dr. Ralph J. Cicerone and Dr. Charles M. Vest are chair and vice chair, respectively, of the National Research Council.

www.national-academies.org

COMMITTEE ON MODELS IN THE REGULATORY DECISION PROCESS

Members

CHRIS G. WHIPPLE (*Chair*), ENVIRON, Inc., Emeryville, CA
M. BRUCE BECK, University of Georgia, Warnell School of Forest Resources, Athens
CLAYTON J. CLARK II, University of Florida, Gainesville
ROBERT T. CLEMEN, Duke University, Fuqua School of Business, Durham, NC
JUDITH A. GRAHAM, American Chemistry Council, Arlington, VA
LOUIS J. GROSS, University of Tennessee, Institute for Environmental Modeling, Knoxville
WINSTON HARRINGTON, Resources for the Future, Washington, DC
PHILIP HOWARD, Syracuse Research Corporation, Environmental Science Center, Syracuse, NY
KIMBERLY L. JONES, Howard University, Washington, DC
THOMAS E. MCKONE, University of California, Berkeley, and the Lawrence Berkeley National Laboratory, Berkeley, CA
NAOMI ORESKES, University of California, San Diego
SPYROS N. PANDIS, University of Patras, Patra, Greece
LOUISE M. RYAN, Harvard School of Public Health, Department of Biostatistics, Boston, MA
MICHAEL L. STEIN, University of Chicago, Statistics Department, Chicago, IL
WENDY E. WAGNER, University of Texas School of Law, Austin

Staff

K. JOHN HOLMES, Project Director
MATTHEW RUSSELL, Associate Staff Officer
RUTH E. CROSSGROVE, Senior Editor
MIRSADA KARALIC-LONCAREVIC, Manager, Technical Information Center
RADIAH A. ROSE, Senior Editorial Assistant

Sponsor

U.S. ENVIRONMENTAL PROTECTION AGENCY
U.S. DEPARTMENT OF TRANSPORTATION

BOARD ON ENVIRONMENTAL STUDIES AND TOXICOLOGY[1]

Members

Senior Staff

---

[1]This study was planned, overseen, and supported by the Board on Environmental Studies and Toxicology.

OTHER REPORTS OF THE BOARD ON ENVIRONMENTAL STUDIES AND TOXICOLOGY

Toxicity Testing in the Twenty-first Century: A Vision and a Strategy (2007)
Sediment Dredging at Superfund Megasites: Assessing the Effectiveness (2007)
Environmental Impacts of Wind-Energy Projects (2007)
Scientific Review of the Proposed Risk Assessment Bulletin from the Office of Management and Budget (2007)
Assessing the Human Health Risks of Trichloroethylene: Key Scientific Issues (2006)
New Source Review for Stationary Sources of Air Pollution (2006)
Human Biomonitoring for Environmental Chemicals (2006)
Health Risks from Dioxin and Related Compounds: Evaluation of the EPA Reassessment (2006)
Fluoride in Drinking Water: A Scientific Review of EPA's Standards (2006)
State and Federal Standards for Mobile-Source Emissions (2006)
Superfund and Mining Megasites—Lessons from the Coeur d'Alene River Basin (2005)
Health Implications of Perchlorate Ingestion (2005)
Air Quality Management in the United States (2004)
Endangered and Threatened Species of the Platte River (2004)
Atlantic Salmon in Maine (2004)
Endangered and Threatened Fishes in the Klamath River Basin (2004)
Cumulative Environmental Effects of Alaska North Slope Oil and Gas Development (2003)
Estimating the Public Health Benefits of Proposed Air Pollution Regulations (2002)
Biosolids Applied to Land: Advancing Standards and Practices (2002)
The Airliner Cabin Environment and Health of Passengers and Crew (2002)
Arsenic in Drinking Water: 2001 Update (2001)
Evaluating Vehicle Emissions Inspection and Maintenance Programs (2001)
Compensating for Wetland Losses Under the Clean Water Act (2001)
A Risk-Management Strategy for PCB-Contaminated Sediments (2001)
Acute Exposure Guideline Levels for Selected Airborne Chemicals (five volumes, 2000-2007)
Toxicological Effects of Methylmercury (2000)
Strengthening Science at the U.S. Environmental Protection Agency (2000)
Scientific Frontiers in Developmental Toxicology and Risk Assessment (2000)
Ecological Indicators for the Nation (2000)
Waste Incineration and Public Health (2000)
Hormonally Active Agents in the Environment (1999)
Research Priorities for Airborne Particulate Matter (four volumes, 1998-2004)
The National Research Council's Committee on Toxicology: The First 50 Years (1997)
Carcinogens and Anticarcinogens in the Human Diet (1996)
Upstream: Salmon and Society in the Pacific Northwest (1996)
Science and the Endangered Species Act (1995)
Wetlands: Characteristics and Boundaries (1995)
Biologic Markers (five volumes, 1989-1995)
Review of EPA's Environmental Monitoring and Assessment Program (three volumes, 1994-1995)
Science and Judgment in Risk Assessment (1994)
Pesticides in the Diets of Infants and Children (1993)

Dolphins and the Tuna Industry (1992)
Science and the National Parks (1992)
Human Exposure Assessment for Airborne Pollutants (1991)
Rethinking the Ozone Problem in Urban and Regional Air Pollution (1991)
Decline of the Sea Turtles (1990)

Copies of these reports may be ordered from the National Academies Press
(800) 624-6242 or (202) 334-3313
www.nap.edu

PREFACE

The use of computational models is an essential element of the environmental regulatory process. The complex relationship between environmental emissions, the quality of the environment, and human and ecological impacts are linked by modeling in the regulatory process. The U.S. Environmental Protection Agency (EPA) may make a scientific determination of basic environmental goals, such as how clean our air and water need to be to protect human health and the environment. But determining how those goals can be met while simultaneously allowing for basic economic services, such as transportation, energy, and agriculture, requires that we examine the links, for example, between the auto emission standards and the attainment of ambient air quality standards or between the point sources of water pollution and the quality of water. The spatial and temporal scales on which environmental controls and environmental quality are linked generally do not allow for an observational approach to understand the links between economic activity and environmental quality. These linkages are made by modeling.

The task undertaken by this committee for the National Academies was to assess evolving scientific and technical issues related to the development, selection, and use of computational and statistical models in the regulatory process at EPA. In this report, the committee provides advice concerning management, evaluation, and use of models at the agency. Through public workshops and other means, the committee has considered cross-discipline issues related to model development and use, performance evaluation, peer review, uncertainty, and quality assurance–quality control. The committee assessed scientific and technical criteria that should be considered in deciding whether a model and its results could serve as a reasonable basis for environmental regulatory activities. It also examined case studies of model development, evaluation, and application as a basis for arriving at guiding principles.

This report has been reviewed in draft form by persons chosen for their diverse perspectives and technical expertise in accordance with procedures approved by National Research Council (NRC) Report Review Committee. The purpose of this independent review is to provide candid and critical comments that will assist the institution in making its published report as sound as possible and to ensure that the report meets institutional standards of objectivity, evidence, and responsiveness to the study charge. The review comments and draft manuscript remain confidential to protect the integrity of the deliberative process. We wish to thank the following for their review of this report: George V. Alexeeff, California EPA; Eula Bingham, University of Cincinnati; John Bredehoeft, the Hydrodynamics Group; E. Donald Elliott, Willkie, Farr & Gallagher, LLP; Paul Gilman, Oak Ridge Center for Advanced Studies; James Hammitt, Harvard Center for Risk Analysis; Michael Koerber, Lake Michigan Air Directors Consortium; Charles Lucas, American International Group, Inc. (retired); Virginia McConnell, Resources for the Future, Inc.; Jana Milford, University of Colorado and Environmental Defense; Lee Mulkey, University of Georgia; Kenneth Reckhow, Duke University; and Scott Zeger, Johns Hopkins University.

Although the reviewers listed above have provided many constructive comments and suggestions, they were not asked to endorse the conclusions or recommendations, nor did they see the final draft of the report before its release. The review of this report was overseen by John Bailar, University of Chicago (retired), and David Allen, University of Texas. Appointed by the NRC, they were responsible for making certain that an independent examination of this report was carried out in accordance with institutional procedures and that all review comments were carefully considered. Responsibility for the final content of this report rests entirely with the committee and the institution.

The committee received oral and written presentations from the following individuals:

Gary Foley, U.S. Environmental Protection Agency
Tom Voltaggio, U.S. Environmental Protection Agency
Albert McGartland, U.S. Environmental Protection Agency
S.T. Rao, U.S. Environmental Protection Agency
Joseph Merenda, U.S. Environmental Protection Agency
Jim Weaver, U.S. Environmental Protection Agency
David Burden, U.S. Environmental Protection Agency
Tim Wool, U.S. Environmental Protection Agency
Leslie Shoemaker, Tetra Tech, Inc.
Jim George, Maryland Department of the Environment
Cecilia Ho, Federal Highway Administration
Harry Kitch, U.S. Environmental Protection Agency
Timothy Miller, U.S. Geological Survey
Jennifer Sass, Natural Resources Defense Council
Scott Slaughter, Center for Regulatory Effectiveness
Adam Finkel, University of Medicine and Dentistry of New Jersey and Princeton University
Gene Tierney, U.S. Environmental Protection Agency
H. Christopher Frey, North Carolina State University
Margo Schwab, Office of Management and Budget
John Graham, Rand Graduate School
Rob Howard, Bechtel-SAIC, LLC
Sheila Jasanoff, Harvard University
Daniel Krewski, McLaughlin Centre for Population Health Risk Assessment
Jan M. Zielinski, McLaughlin Centre for Population Health Risk Assessment
Tim Ramsay, McLaughlin Centre for Population Health Risk Assessment
Richard T. Burnett, McLaughlin Centre for Population Health Risk Assessment
George Leavesley, U.S. Geological Survey
Sam Napolitano, U.S. Environmental Protection Agency
Elliot Lieberman, U.S. Environmental Protection Agency
M. Granger Morgan, Carnegie Mellon University
Pasky Pascual, U.S. Environmental Protection Agency
Barbara Petersen, Exponent, Inc. and Durango Software, LLC
James D. Schaub, U.S. Department of Agriculture
Woodrow Setzer, U.S. Environmental Protection Agency
Harvey Clewell, Centers for Health Research
Rory Conolly, U.S. Environmental Protection Agency
Richard Morgenstern, Resources for the Future
Robert Perciasepe, Audubon Society
Kenneth Reckhow, Duke University
Paul Gilman, Oak Ridge Center for Advanced Studies

The committee and I thank all of these individuals for their contributions. A complete list of dates, titles, and presenter names can be found in Appendix B.

The committee and I are also grateful for the assistance of the NRC staff in the preparation of this report. K. John Holmes played a key role in preparing this report as project director. We also thank Raymond Wassel, senior program director of environmental sciences and engineering in the Board on Environmental Studies and Toxicology (BEST), and the other staff members contributing to this report: James Reisa, director of BEST; Steven Gibb, program officer for strategic communications; Ruth Crossgrove, senior editor; Matthew Russell, associate staff officer; Mirsada Karalic-Loncarevic, manager of the Technical Information Center; and Radiah Rose, senior editorial assistant.

As chair, I thank all the members of the committee for their expertise and dedicated effort throughout the study.

Chris Whipple, *Chair*
Committee on Models in the Regulatory Decision Process

CONTENTS

TABLES

BOXES

Models in Environmental Regulatory Decision Making

Summary

Many regulations issued by the U.S. Environmental Protection Agency (EPA) are based on results from computer models. EPA is a global leader in advancing and using models in the environmental regulatory decision process. Yet the agency has not sufficiently leveraged opportunities to improve its regulatory decisions by adopting a comprehensive strategy for periodically evaluating and refining its models. This report recommends a series of guidelines and principles that, if adopted, will improve environmental regulatory models and decisions made by the agency. Moreover, adoption of these principles will enhance the agency's ability to respond to recent information-quality requirements by allowing EPA to provide more informed responses to outside challenges and reduce the likelihood of erroneous data releases that can prompt challenges.

Models have a long history of helping to explain scientific phenomena and of predicting outcomes and behavior in settings where empirical observations are limited or not available. The use of models has resulted in great advances in scientific understanding and in improvements in a wide array of endeavors. However, by their very nature, all models are simplifications and approximations of the real world. Complex relationships are often simplified, and relationships viewed as unimportant are sometimes eliminated from consideration to reduce computational difficulties and increase transparency.

This report looks specifically at the use of computational models in environmental regulatory activities, particularly at EPA. The use of computational models is central to the regulatory decision-making process because the agency must do prospective analyses of its policies, including estimating possible future effects on the environment, human health, and the economy. Obtaining a comprehensive set of measurement data is not feasible in many cases because of time and resource constraints. The agency uses models to generate estimates (or predictions) when data are not available. EPA also uses models to analyze measurement data for trends and effects. The results of models can become the basis for such decisions as initiating environmental cleanup or regulation. In sum, models are critical tools that help to inform and set priorities in environmental policy development, implementation, and evaluation at EPA.

Because of the critical role played by models, EPA has developed a variety of policies and programs to improve models and their use at the agency. One laudable step has been the establishment of the Council for Regulatory Environmental Modeling (CREM) in 2000 to support modeling activities across the agency and to provide an important resource for interested parties outside of EPA.

The National Research Council (NRC) convened the Committee on Models in the Regulatory Decision Process in response to a request from CREM to independently assess evolving scientific and technical issues related to the selection and use of computational and statistical models in decision-making processes at EPA. The full charge is provided in Box S-1 at the end of the Summary.

MODEL USE IN THE REGULATORY PROCESS AT EPA

Models will always be constrained by computational limitations, assumptions, and knowledge gaps. They can best be viewed as tools to help inform decisions rather than as machines to generate truth or make decisions. Scientific advances will never make it possible to build a perfect model that accounts for every aspect of reality or to prove that a given model is correct in all respects for a particular regulatory application. These characteristics make evaluation of a regulatory model more complex than solely a comparison of measurement data with model results. They suggest that model evaluation be viewed as an integral and ongoing part of the life cycle of a model, from problem formulation and model conceptualization to the development and application of a computational tool. Evaluation of regulatory models also must address a more complex set of trade-offs than evaluation of research models for the same class of models. Regulatory model evaluation must consider how accurately a particular model application represents the system of interest while being reproducible, transparent, and useful for the regulatory decision at hand. Meeting these needs may require different forms of peer review, uncertainty analysis, and extrapolation methods. It also implies that regulatory models should be managed in a way to enhance models in a timely manner and assist users and others to understand a model's conceptual basis, assumptions, input data requirements, and life history.

EPA has played a major role in advancing the science of environmental modeling. However, as with virtually any component of regulatory decision making, improvements to EPA's efforts are possible. Many of the recommendations in this report are derived from a review of current modeling practices within individual EPA research and program offices. This report aims to provide an across-agency vision for the use of models in the future. In keeping with the study charge, the report provides a set of guidelines for improving the use of models to support regulation. The committee offers recommendations in three areas of the modeling process: (1) model evaluation; (2) principles for model development, selection, and application; and (3) model management.

MODEL EVALUATION

Life-Cycle Model Evaluation

Models begin their life cycle with the identification of a need and the development of a conceptual approach, and proceed through building of a computational model and subsequent applications. Models also can evolve through multiple versions that reflect new scientific findings, acquisition of data, and improved algorithms. Model evaluation is the process of deciding whether and when a model is suitable for its intended purpose. This process is not a strict validation or verification procedure but is one that builds confidence in model applications and increases the understanding of model strengths and limitations. Model evaluation is a multifaceted activity involving peer review, corroboration of results with data and other information, quality assurance and quality control checks, uncertainty and sensitivity analyses, and other activities. Even when a model has been thoroughly evaluated, new scientific findings may raise unanticipated questions, or new applications may not be scientifically consistent with the model's intended purpose.

Recommendations

Evaluation of a regulatory model should continue throughout the life of a model. In particular, model evaluation should not stop with the evaluation activities that often occur before the public release of a model but should continue throughout regulatory applications and revisions to the model. For all models used in the regulatory process, the agency should begin by developing a life-cycle model evaluation plan commensurate with the regulatory application of the model (for example, the scientific complexity, the precedent-setting potential of the modeling approach or application, the extent to which previous evaluations are still applicable, and the projected impacts of the associated regulatory decision). Some plans may be brief, whereas other plans would be extensive. At a minimum each plan should

- Describe the model and its intended uses.
- Describe the relationship of the model to data, including the data for both inputs and corroboration.
- Describe how such data and other sources of information will be used to assess the ability of the model to meet its intended task.
- Describe all the elements of the evaluation plan by using an outline or diagram showing how the elements relate to the model's life cycle.
- Describe the factors or events that might trigger the need for major model revisions or the circumstances that might prompt users to seek an alternative model. These could be fairly broad and qualitative.
- Identify responsibilities, accountabilities, and resources needed to ensure implementation of the evaluation plan.

It is essential that the agency is committed to the concept that model evaluation continues throughout a model's life. Model evaluation should not be an end unto itself but a means to an end, namely, a model fitted to its purpose. EPA should develop a mechanism that oversees the evaluation process to ensure that an evaluation plan is developed, resources are committed to carry it out, and modelers respond to what is learned. Although the committee does not make organizational recommendations or recommendations on the level of effort that should be expended on any particular type of evaluation, it recognizes that the resource implications for implementing life-cycle model evaluation are potentially substantial. However, given the importance of modeling activities in the regulatory process, such investments are critical to enable environmental regulatory modeling to meet challenges now and in the future.

Peer Review

Peer review is an important tool for improving the quality of scientific products and is basic to all stages of model evaluation. One-time reviews, of the kind used for research articles published in the literature, are insufficient for many of the models used in the environmental regulatory process. More time, effort, and variety of expertise are required to conduct and respond to peer review at different stages of the life cycle, especially for complex models.

Recommendations

Peer review should be considered, but not necessarily performed, at each stage in a model's life cycle. Some simple, uncontroversial models might not require any peer review, whereas others might merit peer review at several stages. Appropriate peer review requires an effort commensurate with the complexity and significance of the model application. When a model peer review is undertaken, EPA should allow sufficient time, resources, and structure to assure an adequate review. Reviewers should receive not only copies of the model and its documentation but also documentation of its origin and history. Peer review for some regulatory models should involve comparing the model results with known test cases, reviewing the model code and documentation, and running the model for several types of problems for which the model might be used. Reviewing model documentation and results is not sufficient peer review for many regulatory models.

Because many stakeholders and others interested in the regulatory process do not have the capability or resources for a scientific peer review, they need to be able to have confidence in the evaluation process. This need requires a transparent peer review process and continued adherence to criteria provided in EPA's guidance on peer review. Documentation of all peer reviews, as well as evidence of the agency's consideration of comments in developing revisions, should be part of the model origin and history.

Quantifying and Communicating Uncertainty

There are two critical but distinct issues in uncertainty analysis for regulatory environmental modeling: what kinds of analyses should be done to quantify uncertainty, and how these uncertainties should be communicated to policy makers.

Quantifying Uncertainty

A wide range of possibilities is available for performing model uncertainty analysis. At one extreme, all model uncertainties could be represented probabilistically, and the probability distribution of any model outcome of interest could be calculated. However, in assessing environmental regulatory issues, these analyses generally would be quite complicated to carry out convincingly, especially when some of the uncertainties in critical parameters have broad ranges or when the parameter uncertainties are difficult to quantify. Thus, although probabilistic uncertainty analysis is an important tool, requiring EPA to do complete probabilistic regulatory analyses on a routine basis would probably result in superficial treatments of many sources of uncertainty. The practical problems of performing a complete probabilistic analysis stem from models that have large numbers of parameters whose uncertainties must be estimated in a cursory fashion. Such problems are compounded when models are linked into a highly complex system, for example, when emissions and meteorological model results are used as inputs into an air quality model.

At the other extreme, scenario assessment and/or sensitivity analysis could be used. Neither one in its simplest form makes explicit use of probability. For example, a scenario assessment might consider model results for a relatively small number of plausible cases (for example, "pessimistic," "neutral," and "optimistic" scenarios). Such a deterministic approach is easy to implement and understand. However, scenario assessment does not typically include information corresponding to conditions not included in the assessment and whatever is known about each scenario's likelihood.

It is not necessary to choose between purely probabilistic approaches and deterministic approaches. Hybrid analyses combining aspects of probabilistic and deterministic approaches might provide the best solution for quantifying uncertainties, given the finite resources available for any analysis. For example, a sensitivity analysis might be used to determine which model parameters are most likely to have the largest impacts on the conclusions, and then a probabilistic analysis could be used to quantify bounds on the conclusions due to uncertainties in those parameters. In another example, probabilistic methods might be chosen to quantify uncertainties in environmental characteristics and expected human health impacts, and several plausible scenarios might be used to describe the monetization of the health benefits.

Questions about which of several plausible models to use can sometimes be the dominant source of uncertainty and, in principle, can be handled probabilistically. However, a scenario assessment approach is particularly appropriate for showing how different models yield differing results.

Communicating Uncertainties

Effective decision making will require providing policy makers with more than a single probability distribution for a model result (and certainly more than just a single number, such as the expected net benefit, with no indication of uncertainty). Such summaries obscure the sensitivities of the outcome to individual sources of uncertainty, thus undermining the ability of policy makers to make informed decisions and constraining the efforts of stakeholders to understand the basis for the decisions.

Recommendations

Quantifying Uncertainty

In some cases, presenting results from a small number of model scenarios will provide an adequate uncertainty analysis (for example, cases in which the stakes are low, modeling resources are limited, or insufficient information is available). In many instances, however, probabilistic methods will be necessary to characterize properly at least some uncertainties and to communicate clearly the overall uncertainties. Although a full Bayesian analysis that incorporates all sources of information is desirable in principle, in practice, it will be necessary to make strategic choices about which sources of uncertainty justify such treatment and which sources are better handled through less formal means, such as consideration of how model outputs change as an input varies through a range of plausible values. In some applications, the main sources of uncertainty will be among models rather than within models, and it will often be critical to address these sources of uncertainty.

Communicating Uncertainty

Probabilistic uncertainty analysis should not be viewed as a means to turn uncertain model outputs into policy recommendations that can be made with certitude. Whether or not a complete probabilistic uncertainty analysis has been done, the committee recommends that various approaches be used to communicate the results of the analysis. These include hybrid approaches in which some unknown quantities are treated probabilistically and others are explored in scenario-assessment mode by decision makers through a range of plausible values. Effective uncertainty communication requires a high level of interaction with the relevant decision makers to ensure that they have the necessary information about the nature and sources of uncertainty and their consequences. Thus, performing uncertainty analysis for environmental regulatory activities requires extensive discussion between analysts and decision makers.

The Interdependence of Models and Measurements

The interdependence of models and measurements is complex and iterative for several reasons. Measurements help to provide the conceptual basis of a model and inform model development, including parameter estimation. Measurements are also a critical tool for corroborating model results. Once developed, models can drive priorities for measurements that ultimately get used in modifying existing models or in developing new ones.

Measurement and model activities are often conducted in isolation. For example, modelers often add details to models without sufficient measurements to justify or confirm the importance of these changes. Likewise, field and laboratory scientists might expand their compilation of samples without understanding the utility of such information for modeling. Although environmental data systems serve a range of purposes, including compliance assessment, monitoring of trends in indicators, and basic research performance, the importance of models in the regulatory process requires measurements and models to be better integrated. Adaptive strategies that rely on iterations of measurements and modeling, such as those discussed in the 2003 NRC report titled *Adaptive Monitoring and Assessment for the Comprehensive Everglades Restoration Plan,* provide examples of how improved coordination might be achieved.

Recommendations

Using adaptive strategies to coordinate data collection and modeling should be a priority of decision makers and those responsible for regulatory model development and application. The interdependence of measurements and modeling needs to be fully considered as early as the conceptual model development phase. Developing adaptive strategies will benefit from the contributions of modelers, measurement experts, decision makers, and resource managers.

Retrospective Analysis of Models

EPA has been involved in the development and application of computational models for environmental regulatory purposes for as long as the agency has been in existence. Its reliance on models has only increased over time. However, attempts to learn from prior experiences with models and to apply these lessons have been insufficient.

Recommendations

The committee recommends that EPA conduct and document the results of retrospective reviews of regulatory models not only on single models but also at the scale of model classes, such as models of groundwater flow and models of health risks. The goal of such retrospective evaluations should be the identification of priorities for improving regulatory models. One objective of this analysis would be to investigate systematic strengths and weaknesses that are characteristic of various types of models. A second important objective would be to study the processes (for example, approaches to model development and evaluation) that led to successful models and model applications.

In carrying out a retrospective analysis, it might be helpful to use models or categories of models that are old by current modeling standards, because the older models could present the best opportunities to assess actual model performance quantitatively by using subsequent advances in modeling and in new observations.

PRINCIPLES FOR MODEL DEVELOPMENT, SELECTION, AND APPLICATION

Model Parsimony

Models are always incomplete, and efforts to make them more complete can be problematic. As features and capabilities are added to a model, the cumulative effect on model performance needs to be evaluated carefully. Increasing the complexity of models without adequate consideration can introduce more model parameters with uncertain values, and decrease the potential for a model to be transparent and accessible to users and reviewers. It is often preferable to omit capabilities that do not improve model performance substantially. Even more problematic are models that accrue substantial uncertainties because they contain more parameters than can be estimated or calibrated with available observations.

Recommendations

Models used in the regulatory process should be no more complicated than is necessary to inform regulatory decisions. In the process of evaluating whether a model is suitable for its given application, there should be a critical evaluation of whether the model has been made unreasonably complicated. This evaluation should include how model developers and those that select a model for a particular application have addressed the trade-offs between the need for a given model application to be an accurate representation of the system of interest and the need for it to be reproducible, transparent, and useful for the regulatory decision at hand.

Extrapolation

Model use in the environmental regulatory process may involve using the model to extrapolate beyond conditions for which the model was constructed or calibrated or conditions for which the model outputs cannot be verified. For example, it might be necessary to extrapolate laboratory animal data to assessments of possible human effects or to extrapolate the recent history of global environmental conditions to future conditions. In these circumstances, uncertainties about the form of a model and the parameters in the model might yield large uncertainties in model outputs. This problem can be compounded by making a model more complex if the additional processes in the more complex model are unimportant; any extra parameters that need to be estimated could degrade the confidence in the estimates of all parameters.

Recommendations

Extrapolating far beyond the available data for the model draws particular attention in the evaluation process to the theoretical basis of the model, the processes represented in the model, and the parameter values. When critical model parameters are estimated largely on the basis of matching model output to historical data, care must be taken to provide uncertainty estimates for the extrapolations, especially for models with many uncertain parameters.

Proprietary Models

A model is proprietary if any component that is a fundamental part of the model's structure or functionality is not available for free to the general public. The use of proprietary models in the regulatory process can produce distrust among regulated parties and other interested individuals and groups because their use might prevent those affected by a regulatory decision from having access to a model that may have affected the decision. There are many ways in which a model can be proprietary, and some are more prone to engender distrust than others. For example, a model that uses proprietary algorithms may cause more concern than a model that uses publicly available algorithms but has a proprietary user interface.

Recommendations

The committee recommends that EPA adopt a preference for nonproprietary software for environmental modeling. When developing a model, EPA should establish and pursue a goal of not using proprietary elements. It should only adopt proprietary models when a clear and well-documented case has been made that the advantages of using such models outweigh the costs in lower credibility and transparency that accompanies reliance on proprietary models. Furthermore, proprietary models should be subject to rigorous quality requirements and to peer review that is equivalent to peer review for public models. If necessary, nondisclosure agreements could be used for experts to perform a thorough review of the proprietary portions of the model. The review process and results could then be made public without compromising proprietary features. General-purpose proprietary software (for example, Excel, SAS, and MATLAB) usually will not require such scrutiny, although EPA should be cognizant of the costs that obtaining and using such software may impose on interested parties.

MODEL MANAGEMENT

Models and Rule-makings

The sometimes contentious setting in which regulatory models are used may impede EPA's ability to implement some of the recommendations in this report, including the life-cycle evaluation process. Even high-quality models are filled with components that are incomplete and must be updated as new knowledge arises. Yet, those attributes may provide stakeholders with opportunities to mount formal challenges against models that produce outputs that they find undesirable. Requirements such as those in the Information Quality Act may increase the susceptibility of models to challenges because outside parties may file a correction request for information disseminated by agencies.

When a model that informs a regulatory decision has undergone the multilayered review and comment processes, the model tends to remain in place for some time. This inertia is not always ideal: the cumbersome regulatory procedures and the finality of the rules that survive them may be at odds with the dynamic nature of modeling and the goal of improving models in response to experience and scientific advances.

In such an adversarial environment, EPA might perceive that a rigorous life-cycle model evaluation is ill-advised from a legal standpoint. Engaging in this type of rigorous review may expose the model to a greater risk of challenges, at least insofar as the agency's review is made public, because the agency is documenting features of its models that need to be improved. Moreover, revising a model can trigger lengthy administrative notice and comment processes. However, an improved model is less likely to generate erroneous results that could lead to additional challenges, and it better serves the public interest.

Recommendations

It is important that EPA institute best practice standards for the evaluation of regulatory models. Best evaluation practices may be much easier for EPA to implement if its resulting rigorous life-cycle evaluation process is perceived as satisfying regulatory requirements, such as those of the Information Quality Act. However, for an evaluation process to meet the spirit and intent of the Information Quality Act, EPA's evaluation process must include a mechanism for any person to submit information or corrections to a model. Rather than requiring a response within 60 days, as the Information Quality Act does, the evaluation process would involve consideration of that information and response at the appropriate time in the model evaluation process.

To further encourage life-cycle evaluation of models that support federal rule-makings, alternative means of soliciting public comment on model revisions need to be devised. For example, EPA could promulgate a separate rule-making that establishes an agency-wide process for the evaluation and adjustment of models used in its rules. Such a programmatic process would allow the agency to provide adequate opportunities for meaningful public comment at important stages of the evaluation and revision of an individual model, without triggering the need for a separate rule-making for each revision. A more rigorous and formalized evaluation processes for models may result in greater deference to agency models by interested parties and by reviewing courts. Such a response could decrease the extent of model challenges through adversarial processes.

Model Origin and History

Models are developed and applied over many years by participants who enter and exit the process over time. The model origin and history can be lost when individual experiences with a model are not documented and archived. Without an adequate record, a model might be incorrectly applied, or developers might be unable to adapt the model for a new application. Poor historical documentation could also frustrate stakeholders who are interested in understanding a model. Finally, without adequate documentation, EPA might be limited in its ability to justify decisions that were critical to model design, development, or model selection.

Recommendations

As part of the evaluation plan, a documented history of important events regarding the model should be maintained, especially after public release. Each documentation should have the model's origin with such key elements as the identity of the model developer and institution, the decisions on critical model design and development, and the records of software version releases. The model documentation also should have elements in "plain English" to communicate with nontechnical evaluators. An understandable description of the model itself, justifications, limitations, and key peer reviews are especially important for building trust.

The committee recognizes that information relevant to model origins and histories is already being collected by CREM and stored in its model database, which is available on the CREM web site. CREM's database includes over 100 models, although updating of this site has declined in recent years. It provides information on obtaining and running the models and on the models' conceptual bases, scientific details, and results of evaluation studies. One possible way to implement the recommendation for developing and maintaining the model history may be to expand CREM's efforts in this direction. The EPA Science Advisory Board review of CREM contains additional recommendations with regard to specific improvements in CREM's database.

Improving Model Accessibility

Stakeholders and others necessarily play a vital role in EPA's use and evaluation of regulatory models. Differing interpretations of data on risk, environmental trends, and a range of social values mean that a broad array of participants will have a stake in the modeling exercise. As a result, various constituencies and individuals must be able to participate in the modeling process through a variety of activities, such as producing their own model results and commenting on and possibly challenging the legitimacy or accuracy of a model.

EPA faces a number of challenges in making its regulatory models, particularly its complex models, accessible to these diverse interests. Nevertheless, EPA has taken some steps to address accessibility to models, including the CREM database of models. This information enhances the transparency and understandability of models to a wide array of interested participants. Despite these efforts, however, stakeholders and others with limited resources or insufficient technical expertise still face substantial barriers to being able to evaluate EPA's models, comment on important model assumptions, or use the models in their own work.

Recommendations

EPA should place a high priority on ensuring that stakeholders and others have access to models for regulatory decision making. To ensure that its models database contains all actively used models, EPA should continue its support for the intra-agency efforts of CREM. A more formal process may be needed to ensure that CREM's models database is complete and updated with information that is at least equivalent to information provided for models currently contained in the database.

Yet, even with a high-quality models database, EPA should continue to develop initiatives to ensure that its regulatory models are as accessible as possible to the broader public and stakeholder community.

The level of effort should be commensurate with the impact of the model use. It is most important to highlight the critical model assumptions, particularly the conceptual basis for a model and the sources of significant uncertainty. Meaningful stakeholder involvement should be solicited at the model development and model application stages of regulatory activity, when appropriate. EPA could improve model accessibility through a variety of activities, such as requiring an additional interface for each model to help to identify the assumptions and sources of parameters and other uncertainties and providing additional user and stakeholder training.

However, even if full information on a model is available, technical expertise will still be required to judge independently its quality and suitability for regulatory application. Each of these recommendations requires staff time and resources, which may be considerable. Thus, EPA's efforts to enhance opportunities for public participation in any particular case must be balanced against other agency priorities.

The committee anticipates that its recommendations will be met with some resistance because of the potentially substantial resources needed for implementing life-cycle model evaluation. However, given the critical importance of having high-quality models for decision making, such investments are essential if environmental regulatory modeling is to meet challenges now and in the future.

BOX S-1 Task Statement

A National Research Council committee will assess evolving scientific and technical issues related to the selection and use of computational and statistical models in decision-making processes at the Environmental Protection Agency (EPA). The committee will provide advice concerning the development of guidelines and a vision for the selection and use of models at the agency. Through public workshops and other means, the committee will consider cross-discipline issues related to model use, performance evaluation, peer review, uncertainty, and quality assurance/quality control. The committee will assess scientific and technical criteria that should be considered in deciding whether a model and its results could serve as a reasonable basis for environmental regulatory activities. It will also examine case studies of model development, evaluation, and application to further elucidate guiding principles. The objective of the committee will be to provide a report that will serve as a fundamental guide for the selection and use of models in the regulatory process at EPA—the goal is to produce a report on models similar to the NRC's 1983 "Red Book" on risk assessment (*Risk Assessment in the Federal Government: Managing the Process*). As part of its scientific assessment, the committee will need to carefully consider the realities of EPA's regulatory mission so as to provide practical advice on model development and use. The report will avoid an overly prescriptive and stringent set of guidelines and will recognize the need for regulatory and policy decisions in the face of incomplete information and uncertainty. In particular, the committee will not attempt to define a numerical standard for accuracy that all models must attain before they can be used in the decision-making process.

The committee will address the following specific issues:

- What scientific and technical factors should be considered in developing model-acceptability and application criteria that address the needs of EPA, as well as those of interested and affected parties?
- How can the agency provide guidance on procedures for appropriate use, peer review, and evaluation of models that is applicable across the range of interdisciplinary regulatory activities undertaken by the EPA?
- How can issues related to input data quality, model sensitivity, uncertainty, and the use of model outputs be addressed in a unified manner across the multiple disciplines that encompass modeling at EPA?
- Models developed outside of the agency must meet the same acceptability and application criteria as models developed within EPA. How can users of proprietary models meet acceptability and application criteria for the use of models in environmental regulatory applications while maintaining the possible proprietary nature of the code?
- Are there unique evaluation issues associated with different categories of models, such as statistical dose-response models based on epidemiological data?
- How can models be improved in an adaptive management process to allow simpler tools and models to be used now while having the flexibility to incorporate new data, scientific advances, and advances in modeling in the future?
- How can uncertainties and limitations of models be effectively communicated to policy-makers and others who are not experts in the details of the models? How should secondary uses of models be treated, including communication of model uncertainties and limitations?
- What are the emerging scientific and technologic advances that may affect the selection and use of models? Specifically, what are the emerging sources of data (such as remote sensing and other spatially resolved environmental data, and genomic/proteomic data) and developments in information technology for which EPA will need to prepare?

1

Study Background

Models have a long and illustrious history as tools for helping to explain scientific phenomena and for predicting outcomes and behavior in settings where empirical observations may not be available. Fundamentally, all models are simplifications. Complex relationships are reduced, some relationships are unknown, and ones perceived to be unimportant are eliminated from consideration to reduce computational difficulties and to increase transparency. Thus, all models face inherent uncertainties because human and natural systems are always more complex and heterogeneous than can be captured in a model.

This report looks at a specific aspect of computational modeling, the use of environmental models in federal regulatory activities, particularly at the U.S. Environmental Protection Agency (EPA). The use of computational models is central to the decision-making process at EPA because it must do prospective analysis of its policies, including projecting impacts into the future. In addition, obtaining a comprehensive set of measured data to support a decision is typically impracticable in terms of time and resources or is technically and ethically impossible. The agency uses model results to augment and assess measured data. The results of models can become the basis for decisions, such as initiating environmental cleanup and regulation. In sum, models help to inform and set priorities in environmental policy development and implementation at EPA through the ability to evaluate alternative regulations, provide a framework to assess compliance, and summarize available knowledge needed for regulatory decisions.

EARLY ENVIRONMENTAL MODELS

The earliest uses of mathematics to explain the physical world, an important element of environmental models, came in response to the desire to explain and predict the movement of the night sky, the relationship of notes in a musical scale, and other scientific observations (Mahoney 1998; Eagleton 1999; O'Connor and Robertson 2003; Schichl 2004). Later developments of basic conceptual models that helped further the connections of mathematics and modeling to science include the thirteenth century Fibonacci sequences of rabbit population, Paracelsus's connection of dose to disease in the fifteenth century, and the Copernican model of planetary motions in the sixteenth century. The role that mathematics would play in explaining the physical world is evident in the seventeenth century roots of differential calculus, where physical observations of moving objects led to conceptual models of motion, mathematical representations of motions, and finally predictions of locations (Herrmann 1997).

A large expansion in the use of computational models for understanding environmental science and management came in the nineteenth and early twentieth centuries.[1] Mathematical formulations of basic models were developed for many problems, including atmospheric plume motion (Taylor 1915), human dose-response relationship (Crowther 1924), predator-prey relationships (Lotka 1925), and national economy (Tinbergen 1937). An early example of the level of sophistication possible in computational models is Arrhenius's climate model for assessing the greenhouse effect (Arrhenius 1896). Arrhenius's model is a seasonal, spatially disaggregated climate model that relies on a numerical solution to a set of differential equations that represent surface energy balance. The numerical computations required months of hand calculations (Weart 2003), similar to many early numerical models. The computa-

---

[1]The committee decided to use the term "computational model" rather than "mathematical model." These terms are synonymous. The committee considers the the term computational model to be a better descriptor in the era when these models are solved on computers. However, as noted in the text, computational models emerged long before the invention of the digital computer.

tional difficulties associated with such models prompted Lewis Richardson, an early pioneer in the use of computational fluid dynamics in weather modeling, to imagine a "forecast-factory," having thousands of people performing flow calculations directed by a forecast leader coordinating activities with telegraph and colored lights (Fluent Inc. 2006).

Holmes and Wolman (2001) discussed how other model applications during this same era began to spell out the systems-analysis approach to environmental problems that recognizes the interrelationship of physically disparate elements in the environment and the need to understand these relationships through modeling to develop environmental mitigations. A seminal work for understanding the modeling complexity that developed before the invention of digital computers is the Miami Conservancy District flood control project, planned and constructed from 1914 to 1923 (Morgan 1951; Burgess 1979). This project, under the direction of Arthur Morgan, pioneered the use of complex hydrological, economic, and design optimization models coupled with benefit-cost analysis and expert elicitation to quantitatively assess pre- and post-construction conditions of a complex flood control system (Bock 1918; Woodward 1920; Houk 1921; Engineering Staff of the Miami Conservancy District 1922). Morgan and staff used sophisticated computational and graphical techniques to simulate the operation of their flood control design during flood conditions, develop optimizing techniques to increase the project's efficiency, and perform a detailed economic appraisal of the project's impact on more than 77,000 individual properties.

TRENDS IN ENVIRONMENTAL REGULATORY MODEL USE

The past 25 years has seen a vast increase in the number, variety, and complexity of computational models available for regulatory purposes at EPA. Models have increased in capabilities and sophistication through advances in computer technology, data availability, developer creativity, and increased understanding of environmental processes. Demand for models expanded as the participants in regulatory processes, Congress, EPA, Office of Management and Budget (OMB), stakeholders, and the general public required improved analysis of environmental issues and the consequences of proposed regulations. Demands also increased as policy makers have attempted to improve the ability of environmental regulatory activities to achieve the desired environmental benefits and reduce implementation costs. Individual histories are complex, and regulatory model use in specific fields is tied to specific regulatory and scientific developments. However, regulatory needs and model capabilities are often not aligned perfectly. Box 1-1 briefly describes the history of ozone air quality modeling, one area with a lengthy modeling and regulatory history, and the uneven interactions between policy and science.

While the demand for models has grown, the conceptualization of what a model is has shifted in recent years, especially among those closest to the modeling process. Models are viewed less as truth-generating machines and much more as tools designed to fulfill specific tasks and purposes (Beck et al. 1997). As tools, models serve in the decision-

BOX 1-1 Ozone Modeling and the Irregular Swings Between Policy and Science

The formation of ozone in the lower atmosphere (troposphere) is an exceedingly complex chemical process involving the interaction of oxides of nitrogen (NO_x), volatile organic compounds (VOCs), sunlight, and dynamic atmospheric processes. The basic chemistry of ozone formation was known in the early 1960s (Leighton 1961). Reduction of ozone concentrations in general requires control of either NO_x or VOC emissions or a combination of both. Due to the nonlinearity of atmospheric chemistry, the selection of the emission-control strategy has traditionally relied on air quality models.

One of the first attempts to include the complexity of atmospheric ozone chemistry in the decision-making process was a simple observations-based model, the so-called Appendix J curve (36 Fed. Reg. 8186 [1971]) (see Figure 1-1). The curve was based on measurements for six U.S. cities where such data were available. Reliable NO_x data were virtually nonexistent at that time. On the basis of the maximum ozone concentrations observed at these cities and their estimated VOC emissions, the curve purported to indicate the percentage of VOC emission reduction required to attain the ozone standard in an urban area as a function of the peak concentration of photochemical oxidants observed in that area. The Appendix J curve was based on the hypothesis that reductions of VOC emissions were the most effective emission-control path, and this conceptual model helped define legislative mandates enacted by Congress that emphasized controlling these emissions.

The next step in modeling complexity was the empirical kinetic modeling approach (EKMA) (Dimitriades 1977). EKMA used the improved uncertainty of chemical mechanisms that were under intense development in

(Continued on next page)

the late 1970s and early 1980s (Atkinson and Lloyd 1984) to simulate the airshed of interest, assuming that it is a well-mixed box. The final result of the modeling was three-dimensional plots of ozone concentrations as a function of VOC and NO_x emissions (Figure 1-2) that could be used for the design of emission-control strategies.

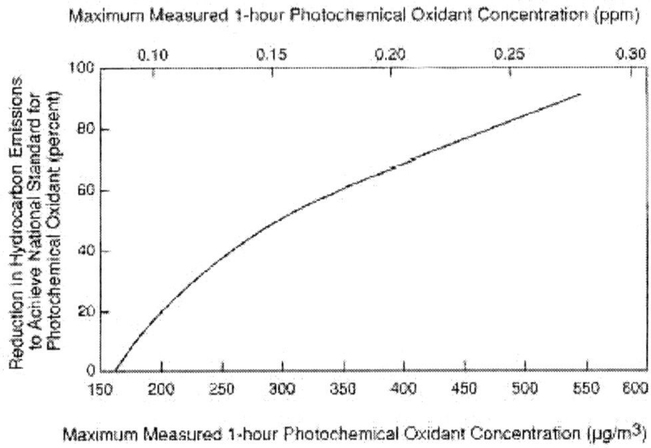

FIGURE 1-1 Appendix J curve. Required hydrocarbon emission control as a function of photochemical oxidant concentration. Source: EPA 1971.

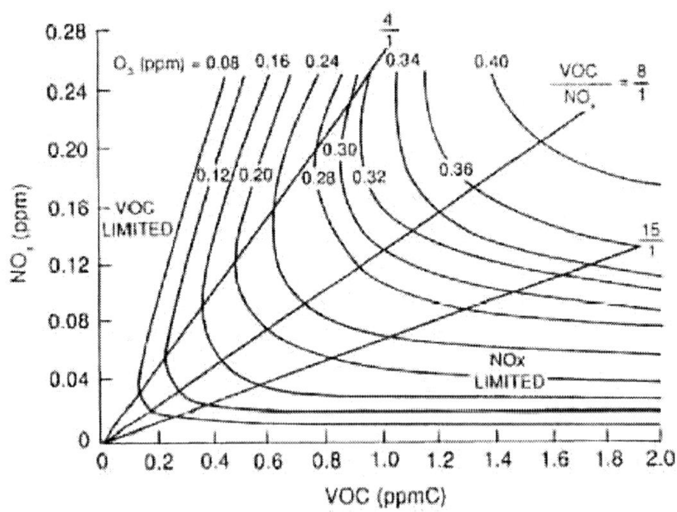

FIGURE 1-2 Typical EKMA diagram. Source: NRC 1991, adapted from Dodge 1977.

The resulting EKMA plots captured the major features and complexities of the NO_x, VOC, and ozone system. For example, they suggested that at low VOC and high NO_x emissions levels, decreases in VOC emissions will reduce peak ozone concentrations, but decreases in NO_x emissions will have the opposite result. Based on the available emissions inventories at the time (1977 to 1982), which turned out to greatly underestimate VOC emissions, many urban areas appeared to be near or above the ridge of the diagram, suggesting that VOC controls were the efficient path. Another characteristic of the EKMA plots is that they suggest that implementation of either VOC or NO_x controls alone is practically always preferable to controlling both ozone precursors. The EKMA approach was heavily used for regulatory applications in the late 1970s and 1980s and supported VOC control as the principal path to attain the ozone standard.

The development of three-dimensional grid models capable of simulating the dynamics and spatial variability of ozone formation (commonly termed 3D chemical transport models or CTMs) also began in the 1970s, although computational demands prevented their use in regulatory activities. EPA in the mid-1970s had committed its research efforts to supporting the development of the urban airshed model (UAM). At the same time, other models (for example, the CIT model) were developed and used by the scientific community (Reynolds et al. 1973). California played a major role in supporting the development and evaluation of these first CTMs. The emphasis of these models was on comprehensive descriptions of the atmospheric system without adjustable parameters (no calibration). During the 1970s, UAM was used only for the Los Angles basin. In the 1980s, the use of 3D models spread to other major metropolitan areas, and the 1990 Clean Air Act Amendments specifically called for the use of such models for all ozone nonattainment areas. The first applications of UAM in the eastern United States also supported the need for VOC controls. Thus, from the early 1970s to the early 1990s, EPA and Congress, with few exceptions, promoted VOC control as the principal path to attaining the ozone standard (for example, required NO_x reductions from motor vehicles).

These VOC reductions had little effect on the ozone concentrations. The incomplete and often erroneous VOC inventories used during this period were one of the major reasons for the choice of suboptimal strategies. For example, biogenic VOC emissions were not included in the inventories until the late 1980s. An influential paper by Chameides et al. (1988) found that when biogenic VOC emissions were included in the inventory in Atlanta and the southeastern United States, NO_x controls were favorable. Additional field and theoretical studies in California suggested that VOC emissions had been underestimated by a factor of approximately two, in large part because of the underestimation of mobile-source emissions. Furthermore, the increased use of regional ozone models that incorporated

(Continued on next page)

long-range transport of ozone and its precursors also demonstrates the importance of NO_x control, especially for regional control of ozone. The debate over a more balanced approach, including control of NO_x emissions, reached a head in the NRC report *Rethinking the Ozone Problem* (NRC 1991; Dennis 2002). The report concluded, "to substantially reduce ozone concentrations in many urban, suburban, and rural areas of the United States, the control of NO_x emissions will probably be necessary in addition to, or instead of, the control of VOCs."

An important aspect of this refocused effort was the need for multistate modeling necessary for addressing transport problems. Although it was originally assumed that ozone problems within a given area were largely caused by emissions within that area, by the end of the 1980s, it was clear that some air quality problems had a larger multistate component and that a substantial contribution to an area's ozone problem could arise from upwind emissions sources. That finding in turn resulted in the formation of multistate organizations, such as the Ozone Transport Commission and Ozone Transport Assessment Group, to develop technical information related to the nature of the transport problem and identify policy options (NRC 2004a). Regional scale modeling is an integral part of understanding the science behind new ambient air quality standards for ozone and fine particulate matter.

There has clearly been a long exchange between policy and science regarding regulations for controlling tropospheric ozone. The choice in the 1970s to concentrate on VOC controls was supported by early results from models. While new results regarding the higher than expected biogenic VOC emissions were being gathered in the 1980s, EPA continued on its path of emphasizing VOC controls, in part because the schedule set by Congress and EPA for attainment of ozone ambient air quality standards was not conducive to reflection on the basic elements of the science (Dennis 2002). The shift in the 1990s toward regulatory activities focusing on NO_x controls from both large stationary sources and mobile sources (along with some VOC controls) was a correction to the prior policy of focusing almost exclusively on VOC reductions. A further complication in the exchange between policy and science during this history was the realization that historical estimates of emissions and the effectiveness of various control strategies in reducing emissions were not accurate. Thus, part of the reason ozone concentrations have not been reduced as much as hoped for over the past 3 decades has been because emissions of some pollutants were much higher than originally estimated and have not been reduced as much as originally predicted. The results of policy decisions to control NO_x takes many years to fully implement, delaying a full understanding of its effectiveness for reducing ozone concentrations. For example, the emissions standards for new on-road diesel engines will not be fully implemented until 2010, and a full fleet turnover will take many years beyond that. While these policies are being implemented, observations of higher weekend ozone when ozone precursor emissions are low (Lawson 2003) and results from an intensive atmospheric observation field campaign in the Houston-Galveston, Texas, area, where highly reactive VOCs seem to play a critical role in ozone formation (Daum et al. 2002), provide new complications to the understanding of the effectiveness of VOC versus NO_x controls.

The long history of the exchange between tropospheric ozone science and modeling and policy demonstrates several critical points. Regulations go forward despite imperfect models and information. The potential harm from environmental hazards can cause regulatory activities to proceed before the science and models are perfected. The long history of controlling VOC and NO_x emissions shows that the inability of the models to predict accurately may reflect not only imperfections in the models but also inputs to the models. In the case of ozone modeling, the inputs to the models (emissions inventories in this case) are often more important than the model science (description of atmospheric transport and chemistry in this case) and require as careful an evaluation as the evaluation of the model. These factors point to the potential synergistic role that measurements play in model development and application. Finally, it is clear that there has been an irregular exchange between modeling/science and policy, which Dennis (2002) describes as "a jerky exchange" between the two, where the policy process has been out of sync with the latest science.

making process as (1) succinctly encoded *archivers* of contemporary knowledge; (2) *interpreters* of links between health and environmental harm from environmental releases to motivate the making of a regulatory decision or policy; (3) *instruments* of analysis and prediction to support the making of a decision or policy; (4) *devices* for communicating scientific notions to a scientifically lay audience; and (5) exploratory *vehicles* for discovery of our ignorance. This committee's task in looking at model use in the regulatory process is 1 and 2, the use of models in understanding environmental impacts and developing and evaluating policy alternatives, that are most prominent. Such analysis of relations and regulatory proposals form the core of regulatory modeling analysis. However, this is not to imply that the other uses of models are not also important for regulatory modeling activities.

It is important to consider why the transition from regarding models as "truth" to regarding models as "tools" might have occurred. Clearly, oversight agencies, such as the OMB, and stakeholders have made an effort to open up the modeling process to external peer review and public scrutiny. As a result, there might be a greater willingness to discuss model shortcomings or at least to disclose them. As regulators become more experienced with the use of models, there might also be a greater appreciation and awareness of the inherent strengths and limitations of models. Finally, the transition to regarding models as tools might represent a push by modelers to educate decision makers that, although models can play an important role in regulatory analysis, models cannot provide "the answer," which is often what the regulatory process demands.

MODEL LIMITATIONS AND ASSUMPTIONS

All models are simplifications of the systems or relations they represent. As a result, the spatial and temporal attributes of processes within a model cannot be resolved fully against observations. Chave and Levin (2003) highlight the intractability of this problem, noting that there is no single correct scale at which to study the dynamics of a natural system. At one end of the spectrum, a model might not simulate at a high enough resolution to represent all critical processes or at scales that capture system heterogeneities. At the other end of the spectrum, an extremely detailed model might not capture large-scale features. These limitations produce two types of uncertainties inherent to models (Morgan 2004). One uncertainty is in the values of key parameters, which are uncertain because of a lack of knowledge and a natural variability. The second uncertainty is in the structure of the model itself. Model uncertainty relates to whether the structure of the model fundamentally represents the system or decision of interest.

These limitations and uncertainties contribute to an inability to ever fully validate or verify numerical models of natural systems (Oreskes et al. 1994). Fundamentally, natural systems are never closed, and model results are never unique. Models of natural systems are never complete, and any match between observations and model results might occur because processes not represented in the model canceled each other. The combination of model formulation and parameters that results in a good match between observations and results is never unique because another combination of model formulation and parameters could result in an equally good match.

In addition, all regulatory model applications have assumptions and default parameters incorporated into them, some of which may include science policy judgments (NRC 1994; EPA 2004a). Assumptions and defaults are unavoidable, as there is never a complete data set to develop a model, but they might have a larger impact on modeling results. Models are commonly used to predict values into the future or under different environmental conditions for which the models were developed, so the assumptions and defaults are subject to debate. Further, the policy settings for regulatory models are framed by more than scientific, technological, and economic ones. Factors related to public values and social and political considerations enter into the modeling process and influence modeling assumptions and defaults.

Although these fundamental uncertainties and limitations are critical to understand when using environmental regulatory models, they do not constitute reasons why modeling should not be performed. When done in a manner that makes effective use of existing science and in a way understandable to stakeholders and the public, models can be very effective for assessing and choosing amongst environmental regulatory activities and communicating with decision makers and the public.

Finally, model results and the observations used to evaluate those results may be at different temporal and/or spatial scales, making it difficult to compare model estimates to actual conditions. For examples, models of climate change, regional groundwater contaminant transport, or human health impacts may make estimates for time scales where observations are not available. Other models, such as air and water quality models, may produce average pollutant concentrations for a wide spatial extent (a grid cell within the model) whereas observations may be available only at a single point within that grid cell.

ORIGIN OF STUDY AND CHARGE TO COMMITTEE

Since the 1980s, EPA recognized the need for agency-wide guidance on the use and development of models, including general model evaluation protocols to test and confirm the accuracy of models. The EPA's Science Advisory Board (SAB), which provides independent scientific and engineering advice to the agency, first issued general guidance on model review in 1989 and recommended that a model's predictive capability could be enhanced through: (1) obtaining external stakeholder input; (2) documenting the model's explicit and implicit assumptions; (3) performing sensitivity analyses; (4) testing model predictions against laboratory and field data; and (5) conducting peer reviews. In 1994 the *Report on the Agency Task Force on Environmental Regulatory Modeling—Guidance, Support Needs, Draft Criteria, and Charter* (EPA 1994a) included guidance for conducting external peer review of models. Other guidance from EPA has come from its Science Policy Council's *Peer Review Handbook* (EPA 2006a) and its National

Center for Environmental Assessment's *Guidelines for Exposure Assessment* (EPA 1992), *Guidelines for Ecological Risk Assessment* (EPA 1998), and *Guidelines for Carcinogenic Risk Assessment* (EPA 2005a).

Despite these efforts to establish and follow appropriate standards, EPA models have become part of the controversies over environmental decision making. At times, Congress has examined models and model results during public hearings, sponsored external reviews of models, or directed EPA to perform a particular analysis (for example, Hearings before the Subcommittee on Oversight and Investigations of the Committee on Commerce, 104th Cong., 1st Sess. 16 [1995]; GAO 1996; NRC 2000, 2001a; EPA 2001a). In addition, models and their results can be prominent in the litigation that results from environmental regulatory activities. EPA has had several environmental regulations overturned because, in the opinion of the courts, the model was considered to be so inaccurate that the regulation was deemed "arbitrary and capricious." McGarity and Wagner (2003) document instances where courts have ruled against the agency because EPA had not sufficiently explained model simplifications, justified the application of a generic model to a specific location, or justified the application of a model to new activities or conditions not originally envisioned when the model was developed. On the other hand, courts have sometimes upheld EPA regulations by ruling in part that EPA's modeling adequately supported their position. In a recent example, the DC Circuit Court of Appeals substantially upheld EPA proposed regulations on "upwind" nitrogen oxides emissions for urban ozone control in part by ruling that the agency's modeling was sufficient to support the determination as to which states should be regulated (D.C. Circuit Court of Appeals, Appalachian Power Co. v. EPA; May 2001).

More recently, the executive branch has been interested in the quality of information produced by government agencies, including EPA. The Office of Management and Budget recently issued guidelines calling for each regulatory agency to develop its own guidance to ensure the quality, objectivity, utility, and integrity of information (OMB 2001). Recognizing the critical roles that models have in developing information, EPA issued information-quality guidelines that include guidance to ensure that the models used in regulatory proceedings be objective, transparent, and reproducible (EPA 2002a). OMB has also issued guidance on peer review (OMB 2004), which EPA has incorporated into its evaluation of models (EPA 2006a).

To help support modeling activities across the agency, EPA established the Council for Regulatory Environmental Modeling (CREM) in 2000. CREM was established to promote consistency and consensus within the agency on mathematical modeling issues, including modeling guidance, development, and application, and to enhance both internal and external communications on modeling activities. CREM is now focused on helping to generate information to determine whether a model and its analytical results are of a quality sufficient to serve as the basis for a decision (Foley 2004). Specifically, the EPA administrator, tasked CREM with developing a guidance document on the development, assessment and use of environmental models; making publicly accessible an inventory of EPA's most frequently used models; consulting with stakeholders concerning modeling issues; holding regional workshops; and engaging with the National Academy of Sciences to produce a report on the use of environmental and human health models for decision making (EPA 2003a). This report is the response to the last charge. Recognizing the importance of EPA regulatory models in their activities, the U.S. Department of Transportation also participated in this study through additional funding and presentations to the committee.

In 2005, the National Research Council (NRC) established the Committee on Models in the Regulatory Decision Process. The Statement of Task set forth to the committee is as follows:

> A National Research Council committee will assess evolving scientific and technical issues related to the selection and use of computational and statistical models in decision-making processes at EPA. The committee will provide advice concerning the development of guidelines and a vision for the selection and use of models at the agency. Through public workshops and other means, the committee will consider cross-discipline issues related to model use, performance evaluation, peer review, uncertainty, and quality assurance/quality control. The committee will assess scientific and technical criteria that should be considered in deciding whether a model and its results could serve as a reasonable basis for environmental regulatory activities. It will also use case examples of EPA's model development, evaluation, and application practices to further elucidate guiding principles. The objective of the committee will be to provide a report that will serve as a fundamental guide for the selection and use of models in the regulatory process at EPA—the goal is to produce a report on models similar to the NRC's 1983 "Red

Book" on risk assessment (NRC 1983). As part of its scientific assessment, the committee will need to carefully consider the realities of EPA's regulatory mission so as to provide practical advice on model development and use. The report will avoid an overly prescriptive and stringent set of guidelines and will recognize the need for regulatory and policy decisions in the face of incomplete information and uncertainty. In particular, the committee will not attempt to define a numerical standard for accuracy that all models must attain before they can be used in the decision-making process.

The task statement asks the committee to address the following specific issues:

- What scientific and technical factors should be considered in developing model acceptability and application criteria that address the needs of EPA, as well as those of interested and affected parties?
- How can the agency provide guidance on procedures for appropriate use, peer review, and evaluation of models that is applicable across the range of interdisciplinary regulatory activities undertaken by EPA?
- How can issues related to input data quality, model sensitivity, uncertainty, and the use of model outputs be addressed in a unified manner across the multiple disciplines that encompass modeling at EPA?
- Models developed outside of the agency must meet the same acceptability and application criteria as models developed within EPA. How can users of proprietary models meet acceptability and application criteria for the use of models in environmental regulatory applications while maintaining the possible proprietary nature of the code?
- Are there unique evaluation issues associated with different categories of models, such as statistical dose-response models based on epidemiological data?
- How can models be improved in an adaptive management process to allow simpler tools and models to be used now while having the flexibility to incorporate new data, scientific advances, and advances in modeling in the future?
- How can uncertainties and limitations of models be effectively communicated to policy makers and others who are not

experts in the details of the models? How should secondary uses of models be treated, including communication of model uncertainties and limitations?

- What are the emerging scientific and technological advances that may affect the selection and use of models? Specifically, what are the emerging sources of data (such as remote sensing and other spatially resolved environmental data, and genomic/proteomic data) and developments in information technology for which EPA will need to prepare?

COMMITTEE APPROACH TO THE CHARGE

The task statement and the interpretation of the task by the committee required it to review and provide recommendations for a wide array of regulatory modeling activities at EPA. The committee is composed of members from many disciplines. Thus, the committee's expertise and the study charge have led it to provide broad recommendations on guidance and principles for improving the general field of regulatory environmental modeling. When individual modeling efforts are examined in the report, it is for illustrative purposes with respect to the study charge. The committee's approach begins with several fundamental definitions.

Basic Definitions

The committee's charge calls for the study to focus on environmental regulatory models. This is clearly a subset of all models used in science, policy making, and elsewhere. To help differentiate environmental regulatory models from other models, the committee defines four basic terms: model, conceptual model, computational model, and environmental regulatory model.

Recognizing the wide usage of the term in academia, policy making, and elsewhere, the committee defines a model as

> a simplification of reality that is constructed to gain insights into select attributes of a particular physical, biological, economic, or social system. Models can be of many different forms. They can be computational. Computational models include those that express the relationships among components of a system using mathemati-

cal relationships. They can be physical, such as models built to analyze effects of hydrodynamic or aeronautical conditions or to represent landscape topography. They can be empirical, such as statistical models used to relate chemical properties to molecular structures or human dose to health responses. Models also can be analogs, such as when nonhuman species are used to estimate health effects on humans. And they can be conceptual, such as a flow diagram of a natural system showing relationships and flows amongst individual components in the environment, a business model that broadly shows the operations and organization of a business, or a model that includes the relationships among both natural and economic components. The above definitions are not mutually exclusive. For example, a computational model may be developed from conceptual and physical models and an animal analog model can be the basis for an empirical model of human health impacts.

Although models range widely in terms of how they are constructed, models share the common objective of aiding in the understanding of a complex and poorly accessible physical, biological, economic, or social system. Figure 1-3 shows one type of model, a physical model for representing planetary motions. Models help generate information to better understand the relationship among components in a system, to extrapolate the behavior of a system to alternate designs, or to projected future conditions. Figure 1-4 shows a second type of model, an analog model where a white mouse is used as analog for estimating human health impacts. This figure also shows one of the issues that arise from using such a model, the need to extrapolate from the range of exposures for a mouse down to the range of exposures for humans. Although the question of whether a mouse is an appropriate analog model for estimating human health impacts is not part of this study, issues related to the statistical models that are used to extrapolate from mice to humans are part of this study.

The process of building computational, physical, and other models begins with a basic conceptualization of a system. A conceptual model is an abstract representation that provides the general structure of a system and the relationships within the system that are known or hypothesized to be important. Many conceptual models have as a key component a graphical or pictorial representation of the system.

Although the environmental regulatory process typically requires numerical analysis of proposed regulations, the conceptual model provides critical synoptic or summary understanding of the principle factors that influence the effectiveness of policies and, thus, is critical for regulatory analysis. In the context of environmental regulatory model applications, conceptual models are critical for both guiding quantitative analysis and communicating with decision makers, stakeholders, and the interested public.

A subset of all models are those that use measurable variables, numerical inputs and mathematical relationships to produce quantitative outputs. The committee defines a computational model as

> a model that is expressed in formal mathematics using equations, statistical relationships, or a combination of the two. Although values, judgment, and tacit knowledge are inevitably embedded in the structure, assumptions, and default parameters, computational models are inherently quantitative, relating phenomena through mathematical relationships and producing numerical results.

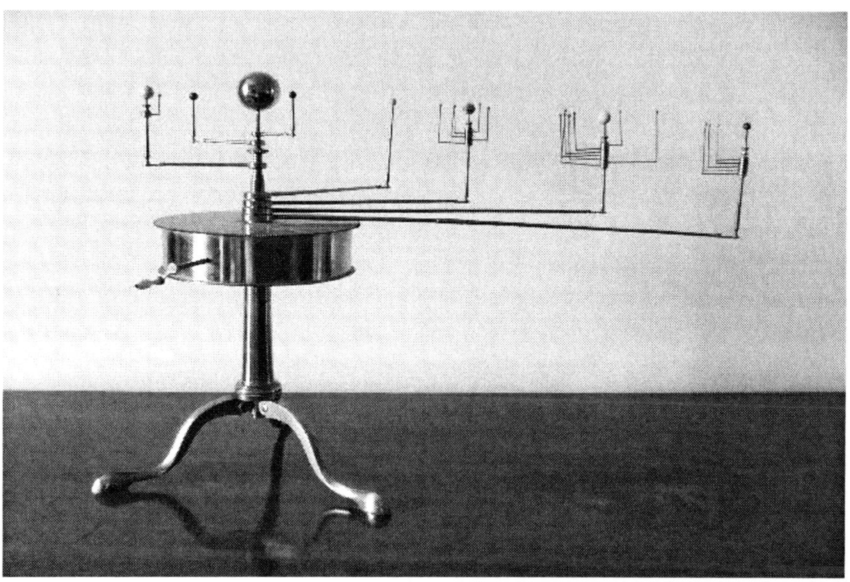

FIGURE 1-3 An orrery or physical model of the solar system. Source: C. Mollan, National Inventory of Scientific Instruments, Royal Dublin Society. Image courtesy of Miruna Popescu. Reprinted with permission; copyright 2004, Armagh Observatory.

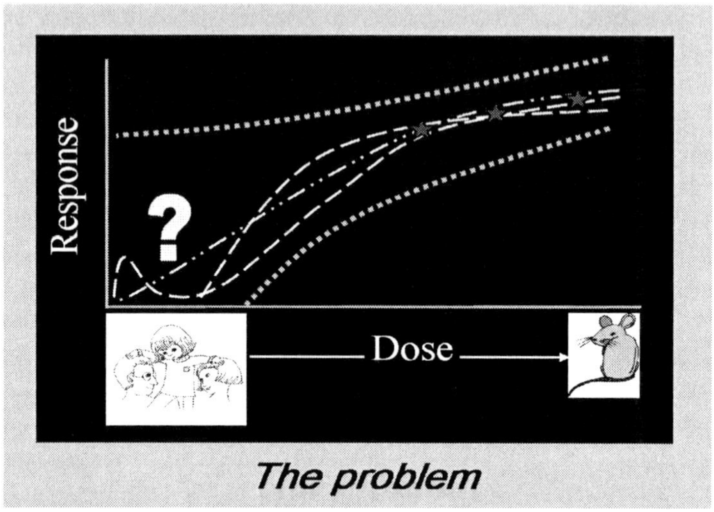

FIGURE 1-4 The use of a mouse model for estimating human health risks. Source: Conolly 2005.

Two examples of computational models are shown in Figures 1-5 and 1-6. Figure 1-5 (taken from Morales et al. 2000) shows the use of statistical models to characterize the lifetime risk of developing bladder cancer among males living in southwestern Taiwan as a function of exposure to arsenic in drinking water (measured in micrograms per liter). Each dot in the three panels represents the estimated lifetime risk for subjects exposed in increments of 100 μg/L, with each panel representing a separate population. We will come back to these figures later in the report, since they provide a very clear illustration of the impact of model choice on estimated dose response. Figure 1-6 shows the conceptual structure of the integrated exposure uptake biokinetic (IEUBK) model that is used to estimate blood lead levels in children. This model has been used in both air quality and hazardous waste-site applications to support standards and cleanup goals (NRC 2005a).

Finally, the committee's task statement concentrates on the application of environmental regulatory models at the EPA. The committee defines an environmental regulatory model as

> a computational model used to inform the environmental regulatory process. Some models are independent of a specific regulation, such as water quality or air quality models that are used in an array

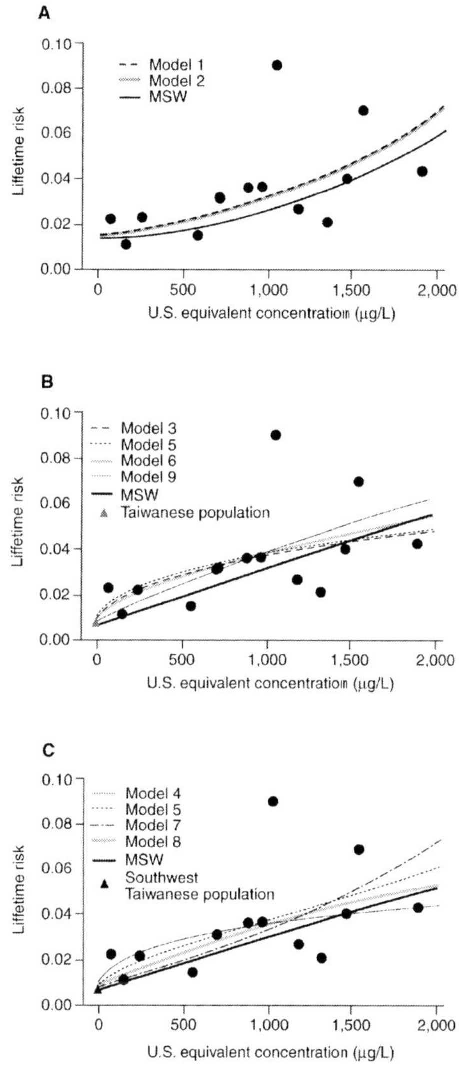

FIGURE 1-5 Examples of dose-response models for estimating lifetime risk for male bladder cancers due to arsenic in drinking water for various exposed populations. A shows the estimated lifetime death risk for male bladder cancer without comparison population; B shows the estimated lifetime death risk for male bladder cancer with Taiwanese-wide comparison population; and C shows the estimated lifetime death risk for male bladder cancer and the southwestern Taiwanese region comparison population. Note that several possible statistical models are fit to each data set. Source: NRC 2001b, from Morales et al. 2000.

of application settings. Other models are created to provide a regulation-specific set of analyses completed during the development and assessment of specific regulatory proposals. The approaches can range from single parameter linear relationship models to models with thousands of separate components and many billions of calculations.

Environmental regulatory models range from those that come complete with source code, documentation, and cellophane packaging to those that are simply a system of algebraic equations or statistical operations. Models also are often coupled together for environmental regulatory applications.

In the context of their use in environmental regulatory activities, the differentiation between a model and its application can be difficult. For example, some models developed for a single set of analysis may be viewed by users as inseparable from their applications, and treated

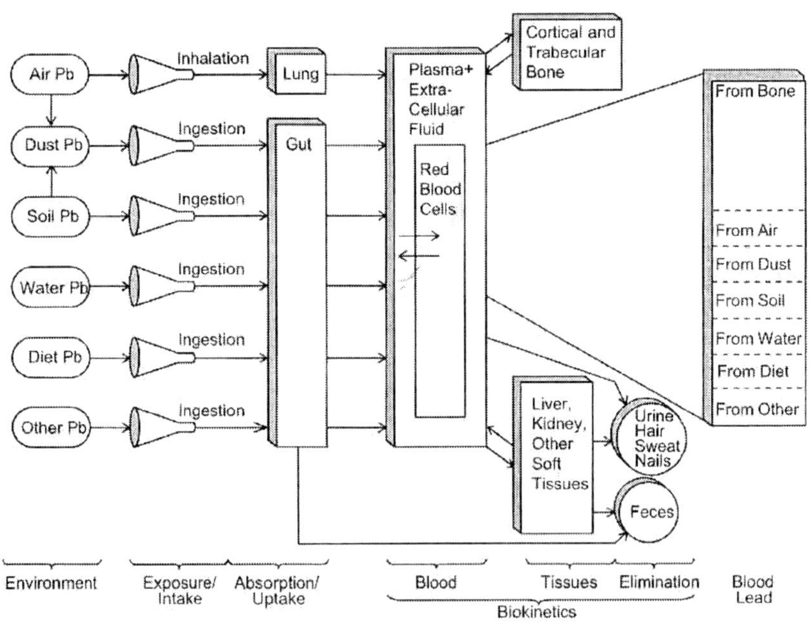

FIGURE 1-6 Components and functional arrangements of the IEUBK model that predict blood lead levels in children. Source: EPA 1994b.

synonymously. In Figure 1-3, each line is from a different model fitted through the same data, resulting in several possible dose-response curves. In other cases, general models are adapted to a particular location or particular contaminant through problem-specific input data or modification of particular assumptions. To further blur the distinction, analysts may use the term "model" to refer to a particular application of a general model.

In some ways this is merely a semantic difference. However, the development and application of a model pose differing evaluation issues. Further, although a model may be applicable to a given setting, its actual application to that setting may be problematic if input parameters for that application are not available or incorrectly specified. Thus, it is necessary to differentiate between evaluation of a general model, evaluation of the applicability of that model to a particular circumstance, and the ultimate implementation of that model, including the specification of parameter and/or input values. This report is not entirely about de novo model development and use, but also the application of previously developed models to specific applications.

What Types of Models Are Within Study Scope

A broad array of environmental models is used in the implementation of EPA's regulatory mission. This includes the use of models in the assessment and regulation of toxic substances, the setting of emissions and environmental standards, and the development of mitigation plans. For example, models are used to

- Assess exposures to contaminants and effects, as well as the relationships between them.
- Project future conditions or trends.
- Extrapolate and interpolate values to situations in which observations are not available.
- Assess the contributions of individual sources to a problem that results from aggregate and/or cumulative exposures.
- Evaluate attributes and impacts of different policy alternatives or future scenarios.
- Evaluating the post-implementation adequacy of a regulation to achieve its goals.

- Consider how the actions of regulated parties might be impacted by alternate policy instruments such as emissions standards versus emissions trading.

The types of models used in this regulatory analysis include those for emissions, environmental fate and transport, exposure, dose (pharmacokinetic models), health effects, ecological impacts, engineering, and economics. Chapter 2 provides more discussion of these models and how they are used in the regulatory process. These models vary widely in complexity. One of the simplest environmental regulatory modeling applications is the use of one-dimensional groundwater flow equations in the assessment of regulatory actions for leaking underground storage tanks (Weaver 2004). Such models use an exact solution to simple differential equations that describe straight-line flow and transport in a homogeneous aquifer. A similar model complexity is a simple linear dose-response model that fits a straight line to a series of individual dose-response points. At the other end of the spectrum is the highly complex Community Multiscale Air Quality (CMAQ) model and its associated meteorological and emissions processing models that simulates the transport, transformation, and formation of multiple atmospheric pollutants. These three models (CMAQ for the simulation of the chemical transformation and fate of pollutants, an emissions model for anthropogenic and natural emissions that are injected into the atmosphere, and a meteorological model for the description of atmospheric states and motion) form a coupled modeling system. CMAQ uses as inputs the results of the emissions and meteorological models. Thus, this suite of individual models can be an even more complex model. Regardless of their level of complexity, all environmental regulatory models provide a quantitative tool for the development, implementation, and assessment of environmental policies.

What Types of Models Are Outside the Study Scope

Although the environmental modeling considered part of the committee's charge encompasses a substantial portion of EPA's modeling activities, some modeling applications are outside of the primary scope. Foremost, the committee is not constituted to comment on the development and use of laboratory animal analog models of human health responses to environmental pollutants. Assessing issues related to the use of an animal species as an analog for human health impacts is outside the expertise of this committee. Another NRC report describes issues related to the use of animal analog models (NRC 2001b). However, computational models, particularly statistical dose-response models, that are used to extrapolate laboratory animal data to humans are included.

Additionally, the committee's focus is on models used in the development, assessment, and implementation of environmental regulatory actions. EPA also uses models in a variety of other applications including planning, project scheduling, data collection, research, prediction, and forecasting. In so far as these models are computational, the committee's recommendations may be useful for these models and their applications. But the committee in no way focused on some of the unique attributes of model selection and use at EPA in these other activities. Because of the wide array of environmental modeling at the agency, there is sometimes not a clear distinction between models used for regulatory purposes that are within the scope of this study and models considered to be used for nonregulatory purposes. For example, the same model may be used for both a regulatory application and a research application. In this way there is sometimes a continuum from models clearly in the regulatory domain under the purview of this study and other applications clearly outside the scope of work. However, not all model applications at EPA directly lead to regulation and there are clearly some model applications that fall outside the committee's scope.

REPORT CONTENTS

This report documents the committee's response to the charge described above. The report consists of six chapters and a summary. Chapter 2 describes the diversity of model use at EPA, how the agency currently integrates models into its policies, and some of the challenges to model use. Chapter 3 discusses the major steps in environmental regulatory model development, focusing on the main lessons learned from previous efforts in EPA. Chapter 4 discusses the evaluation of these models. Chapter 5 describes issues that arise in selecting models for their application in environmental regulatory activities. The report closes by discussing future environmental regulatory model activities in Chapter 6.

2

Model Use in the Environmental Regulatory Decision Process

Regulatory model use at EPA can be contentious. Decisions based on model results might have important public health or environmental consequences and impose substantial costs. Like other aspects of regulation, models are used and evaluated within an environment of legislative requirements, regulatory review, extensive comment by interest groups and other federal agencies, and legal challenge. Within this environment, the development, maintenance, and use of models diverge in important ways from research modeling in academia or nonregulatory modeling in the public and private sectors.

In spite of the challenges, the use of computational models within the regulatory decision process at EPA is a continually growing practice. This growth is in response to greater demands for quantitative assessment of regulatory activities, including analysis of how well environmental regulatory activities fulfill their objectives and at what cost. Models are essential for estimating a variety of relevant characteristics—including pollutant emissions, ambient conditions, and dose—when direct observation would be inaccessible, infeasible, or unethical. Finally, models allow regulators to move away from technology-based regulations that do not use quantitative analysis for assessing their benefits. This chapter describes the diversity of model use at EPA and the current integration of models into its regulatory policies. It highlights how EPA regulatory model use is influenced by legislative mandates and executive orders as well as oversight from the courts and outside participants.

REGULATING WITHOUT COMPUTATIONAL MODELS

Although models are essential tools if regulators are to be able to predict the risks or the effects of their regulations on the natural and human environment, models are neither necessary nor sufficient to produce the regulations themselves. In the 1970s, when the legislative framework underlying most of today's environmental policy was first established, few sophisticated computational environmental models—models designed to predict the environmental consequences of human activity—existed. Moreover, the monitoring networks capable of quantitative description of the state of the environment were rudimentary, and the technology for measurement of pollutant discharges of various kinds and their environmental effects were much less developed than today's technology. It was in this setting that most modern environmental regulatory statutes first appeared, including the Clean Air Act (CAA) of 1967, the Federal Water Pollution Control Act of 1972 and renamed the Clean Water Act (CWA) in 1977, and the Safe Drinking Water Act of 1976. Regulatory designs at the time necessarily minimized the use of computational models in the regulatory process.

The models that did exist played little role in that process because the new environmental statutes emphasized the use of technology-based pollution discharge regulation. Technology-based regulation requires polluters to adopt a particular technology (or, in some cases, achieve a level of performance associated with a particular technology) without regard to the potential or actual environmental improvements that would result.

Even before the implementation of the federal environmental statutes, technology-based regulation partly relied on there being some level of pollution abatement practiced by at least some plants in most industries. EPA was to find those plants and set a performance standard for all plants that was based in some way on what most plants were doing. Usually the congressional mandate involved the use of the words "best technology," and it was left to EPA to interpret and give operational meaning to the various designations of "best." For example, industrial water pollutant dischargers had to meet "best practicable treatment" (BPT) technology standards by 1977 and "best available treatment economically achievable" (BATEA or, more often, BAT) standards by 1983. In industries such as food processing and laundries that generated wastewater that resembled domestic waste (in constituents if not in strength), the usual interpretation of BPT was a performance standard that approximated what good secondary (biological) treatment could do. For other industries, BPT was often defined as the "average of the best" plants in the industry identified as having wastewater treatment in place. For BAT, the standard was the "best of the best," at least until the CWA Amendments of 1977, which redirected BAT toward the control of toxic pollutants in wastewater.

It should be noted that technology-based standards are not the only policy instrument that makes it possible to regulate without having to predict the environmental effects of environmental regulations. Indeed, the need to predict the consequences of regulations depends not on the policy instrument but on the policy goal. If the goal is to achieve a level of emissions reductions rather than environmental quality, there is no need to inquire into environmental effects, regardless of policy instrument. Other environmental policies proposed in the early 1970s shared that property, including several proposals using economic incentives.[1] Like technology-based standards, none of these proposals had an environmental objective beyond the notion that a reduction in effluent discharges would be an improvement and that policies could be fine-tuned later, when scientists had collected more data and achieved a better understanding of environmental processes.

Although the CAA and CWA of the 1970s (as well as other environmental statutes) made extensive use of technology-based standards, it would be misleading to leave the impression that their regulatory arsenals were not limited to such standards. Both statutes also had explicit environmental goals, measurement criteria for determining when the goals were met, and timetables for meeting them. For example, the National Ambient Air Quality Standards (NAAQS) in the 1970 CAA focus on reducing air pollutant concentrations to levels that are protective of human health and public welfare. This legislation required states to develop state implementation plans (SIPs), which are subject to EPA approval. Such approval was contingent on whether the plans, when implemented, would reduce emissions enough to allow the ambient standards to be met. EPA would come to base these SIP approval decisions on emission-inventory models linked to air quality models. In a similar manner, the CWA specified further regulatory action in "water-quality-limited" waters, where the imposition of the technology-based

---

[1]In 1970, a tax on sulfur emissions as a partial alternative to some of the air quality regulation then under consideration in Congress was proposed by President Nixon. In November 1971, an effluent-charge amendment to clean water legislation then under consideration was offered and debated in the Senate (Kelman 1982; Kneese and Schultze 1975).

standards was considered insufficient to achieve water quality standards. Eventually, that section of the CWA gave rise to the "total maximum daily load" program.

Technology-based regulation proved to be a crude approach to pollution abatement policy. Moreover, it did not ultimately relieve Congress and EPA of the need for models to assess whether abatement policies were sufficient to achieve ambient goals. However, at a time when few models were available for linking pollution abatement to environmental improvement, technology-based standards provided a basis for regulating pollutant discharges that did not require knowledge of what the effects of such regulation would be. Today technology-based regulations are still in use, primarily in circumstances in which data and models do not yet permit an adequate assessment of the effects of regulation on environmental or health end points and in which other approaches have failed to generate regulations (these two situations overlap substantially). For example, Title III of the 1990 CAA Amendments changed the primary focus of hazardous air pollutant (HAP) regulation from a risk-based approach to a two-step process, where the primary focus has been on a technology-based approach to mandate promulgation of emissions standards for sources based to some extent on maximum achievable control technologies (MACT), followed by a residual risk assessment. In the preceding regime, regulators made little progress in producing regulations, largely because the inadequacies of data and models linking emissions of HAPs to adverse health effects. The current approach directs EPA to develop a MACT standard for each industrial source category, defined in part by high emissions of listed pollutants. Since 1993, EPA has promulgated over 100 MACT regulations (for the list, see EPA 2006b). After a MACT has been applied, EPA is to perform a residual risk assessment to evaluate the adequacy of the MACT, which might require additional controls if significant risks still exist.

REGULATORY MODEL CLASSIFICATIONS

There are many ways to classify the regulatory models used by EPA, each with its own perspectives and particular advantages and disadvantages. Two broad categorizations are used here: (1) a functional perspective that categorizes models based on their representation of scientific and other processes that translate human activities and natural systems interactions into environmental impacts and (2) a regulatory perspective that categorizes models based on how they are used in environmental regulation. In short, we see these as attempting to represent how an environmental scientist, engineer, or economist might see model use and how a regulator or stakeholder might see model use. In presenting a science perspective and a regulatory perspective, the committee acknowledges that the user community for environmental regulatory models is diverse, and a single perspective on model classification is not possible. More perspectives provide insights into model use, insights that are not possible from a single perspective. Looking at the functions of models as representing different environmental and human processes helps to emphasize the role of individual models and the need to integrate across multiple models for many regulatory activities. Looking at models from the perspective of their role in a complex regulatory setting helps to make clear the role of legislation and regulation in determining modeling objectives and the separate modeling responsibilities for EPA, state, and local governments.

Given the wide range of model applications and large number of models used in environmental regulation, the committee does not attempt to present an inventory of models used by EPA. The most exhaustive inventory with descriptions of individual models is EPA's Council on Regulatory Environmental Models (CREM) (EPA 2006c), although many other web sites are devoted to describing various programs' modeling initiatives (see Table 2-1). CREM's knowledge-base documents more than 100 models used by various offices at the agency. It is the single best, although incomplete, inventory of models at EPA. The information available on each model includes user information on obtaining and running the model and model documentation, including conceptual basis, scientific details, and results of evaluation studies. A full review of the knowledge base has recently been completed by EPA Science Advisory Board and is beyond the committee's charge (EPA 2006d). However, we note that additions to the CREM knowledge base have ceased since 2004, with the exception of several climate change models that were added in 2006. For the knowledge base to reach its full capability, it needs to be updated continually and to include all types of models used at EPA, including those in the health risk assessment field.

Regulatory Models from a Functional Perspective

In this section, we discuss models categorized according to how they fit into a description of the processes that translate human activities and natural systems interactions into environmental impacts. Figure 2-1 shows an illustration of the pathways from activities to emissions to impacts. In the figure, individual components simulate the relationships between human activities and emissions, emissions and concentrations, concentrations and exposures, and exposures and impacts. It also indicates the feedback of impacts on human activities and natural processes. Appendix C provides examples of specific models from the model categories. The figure provides an approximate categorization of how computational models used in environmental analysis have historically been grouped, in particular, in economic, environmental, and human health models. This perspective allows for the identification of particular types of models and the linkages among these models. Each box is highly aggregated and could be expanded into a diagram of sub-boxes. An example of how this aggregate representation might be represented in more detail will be discussed with respect to human health risk assessment in a later section.

The categories of models that are integral to environmental regulation include activity models, natural and anthropogenic emissions models, fate and transport models, exposure models, dose models, human health models, environmental and ecosystem impact models, and economic impact models. Although the categories of models shown in

TABLE 2-1 Examples of EPA's Web Sites Containing Model Descriptions for Individual Programs

National Exposure Research Laboratory Models Web Site http://www.epa.gov/nerl/topics/models.html
Atmospheric Sciences Modeling Division Web Site http://www.epa.gov/asmdnerl/index.html
Office of Water's Water Quality Modeling Web Site http://www.epa.gov/waterscience/wqm
Center for Subsurface Modeling Support Web Site http://www.epa.gov/ada/csmos.html
National Center for Environmental Assessment's Risk Assessment Web Site http://cfpub.epa.gov/ncea/cfm/nceariskassess.cfm?ActType=RiskAssess
National Center for Computational Toxicology Web Site http://www.epa.gov/ncct

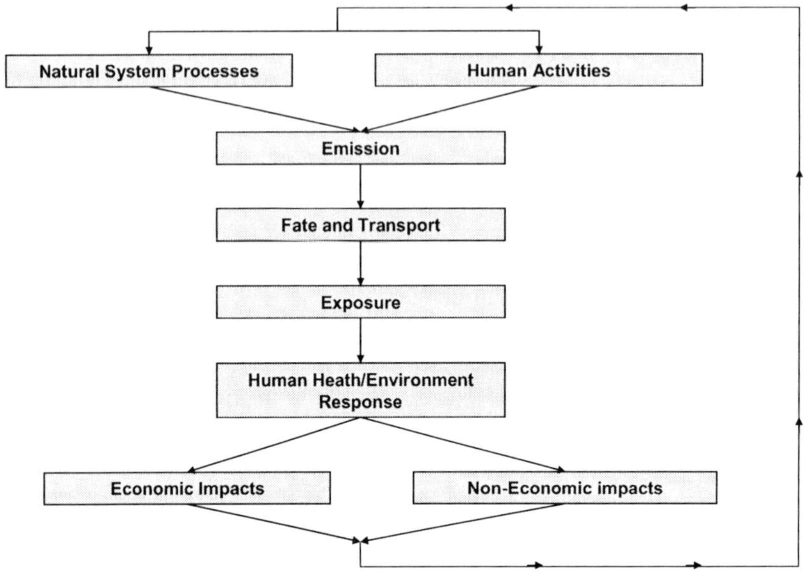

FIGURE 2-1 Basic modeling elements relating human activities and natural systems to environmental impacts.

Figure 2-1 are not specific to environmental media, the models that fit into each category tend to be further subdivided by media. For example, the generic category of environmental fate and transport models can be subdivided further into various types of subsurface containment transport models, surface-water quality models, and air quality models (Schnoor 1996; Ramasawami et al. 2005).

Scope of Regulatory Model Applications

Table 2-2 contains short descriptions of some of EPA's regulatory activities that rely on modeling. These environmental regulatory modeling activities typically occur as a subset of the full system summarized in Figure 2-1. The underlying statutory requirements, the regulations implementing the statutory requirements, and the importance of the activity dictate the nature of the modeling analysis. For example, assessing the toxicity of new pesticides and other chemicals in the environment may focus on just the fate and transport or toxicity portion of the system. Assessing the risks from leaking underground petroleum storage tanks,

TABLE 2-2 Examples of Major EPA Documents That Incorporate a Substantial Amount of Computational Modeling Activities

Air Quality
Criteria Documents and Staff Papers for Establishing NAAQS
Summarize and assess exposures and health impacts for the criteria air pollutants (ozone, particulate matter, carbon monoxide, lead, nitrogen dioxide, and sulfur dioxide). Criteria documents include results from exposure and health modeling studies, focusing on describing exposure-response relationships. For example, the particulate matter criteria document placed emphasis on epidemiological models of morbidity and mortality (EPA 2004c). The Staff Paper takes this scientific foundation a step further by identifying the crucial health information and using exposure modeling to characterize risks that serve as the basis for the staff recommendation of the standards to the EPA Administrator. For example, models of the number of children exercising outdoors during those parts of the day when ozone is elevated had a major influence on decisions about the 8-hour ozone national ambient air quality standard (EPA 1996).
State Implementation Plan (SIP) Amendments
A detailed description of the scientific methods and emissions reduction programs a state will use to carry out its responsibilities under the CAA for complying with NAAQS. A SIP typically relies on results from activity, emissions, and air quality modeling. Model-generated emissions inventories serve as input to regional air quality models and are used to test alternative emission-reduction schemes to see whether they will result in air quality standards being met (e.g., ADEC 2001; TCEQ 2004). Regional scale modeling has become an integral part of developing state implementation plans for new 8-hour ozone and fine particulate matter standards. States, local governments, and their consultants do this analysis.
Regulatory Impact Assessments for Air Quality Rules
RIAs for air quality regulations document the costs and benefits of major emission-control regulations. Recent RIAs have included emissions, air quality, exposure, and health and economic impacts modeling results (e.g., EPA 2004b). See Box 2-3 for a further discussion of the RIA.
Water Regulations
Total Maximum Daily Load (TMDL) Determinations
For each impaired water body, a TMDL documents a state-designated water quality standard need to meet a designated use for that water body and the amount by which pollutant loads need to be reduced to meet the standard. TMDLs utilize water quality and/or nutrient loading models. States and their consultants do the majority of this modeling, with EPA occasionally doing the modeling for particularly contentious TMDLs (EPA 2002b; George 2004; Shoemaker 2004; Wool 2004).
Leaking Underground Storage Tank Program
Assesses the potential risks associated with leaking underground gasoline storage tanks. At an initial screening level, it may assess only one-dimensional transport of a conservative contaminant using an analytical model (Weaver 2004).

(Continued)

TABLE 2-2 Continued

Development of Maximum Contaminant Level for Drinking Water
Assess drinking water standards for public water supply systems. Such assessments can include exposure, epidemiology, and dose-response modeling. (EPA 2002c; NRC 2001b, 2005b).
Pesticides and Toxic Substances Programs
Pre-manufacturing Notice Decisions
Assess risks associated with new manufactured chemicals entering the market. Most chemicals are screened initially as to their environmental and human health risks using structure-activity relationship models.
Pesticide Reassessments
Requires that all existing pesticides undergo a reassessment based on cumulative (from multiple pesticides) and aggregate (exposure from multiple pathways) health risk. This includes the use of pesticide exposure models.
Solid and Hazardous Wastes Regulations
Superfund Site Decision Documents
Includes the remedial investigation, proposed plan, and record of decision documents that detail the characteristics and cleanup of Superfund sites. For many hazardous waste sites, a primary modeling task is utilizing groundwater modeling to assess the movement of toxic substances through the substrate (Burden 2004). The remedial investigation for a mining megasite might include water quality, environmental chemistry, human health risk, and ecological risk assessment modeling (NRC 2005a).
Human Health Risk Assessment
Benchmark Dose (BMD) Technical Guidance Document
EPA relies on both laboratory animal and epidemiologic studies in assessing the noncancer effects of chronic exposure to pollutants (that is, the reference dose [RfD] and the inhalation reference concentration, [RfC]). These data are modeled to estimate the human dose-response. EPA recommends the use of BMD modeling, which essentially fits the experimental data to use as much as the available data as possible (EPA 2000).
Guidelines for Carcinogen Risk Assessment
The cancer guidelines set forth a revised set of recommended principles and procedures to guide EPA scientists and others in assessing the cancer risks resulting from exposure to chemicals or other agents in the environment. One of the principal advancements was to describe approaches that consider mode-of-action data, if available, in the quantitative assessment. The guidelines are also used to inform agency decision makers and the public about risk assessment procedures (EPA 2005a).
Ecological Risk Assessment
Guidelines for Ecological Risk Assessment
The ecological risk assessment guidelines provide general principles and give examples to show how ecological risk assessment can be applied to a wide range of systems, stressors, and biological, spatial, and temporal scales. They describe the strengths and limitations of alternative approaches and emphasize processes and approaches for analyzing data rather than specifying data collection techniques, methods, or models (EPA 1998).

especially during initial assessments, focuses solely on the fate-transport component. The SIP process, which involves extensive emissions and air quality modeling, stops at simulating atmospheric concentrations of air pollutants. Ideally, regulations would be informed by understanding the whole of the paradigm, shown in Figure 2-1, from human activities through adverse outcomes. However, only the most important regulatory assessments, such as some of those done for federal rules that have major economic impacts, include a simulation of processes from activity to health impacts. These are the rules that generate most of the benefits and costs of environmental regulation, and the modeling effort can be enormous. A recent example of such an analysis is the regulatory impact assessment (RIA) for the control of air pollutant emissions from nonroad diesel engines (EPA 2004b). Even the extensive modeling that accompanied this rule cannot quantitatively consider all aspects of the problem. For example, in discussing behavioral responses to increasing costs for nonroad diesel engines, stakeholders suggested that equipment users may substitute different equipment (gasoline engines) or even labor (the use of a laborer and shovel instead of a backhoe) for more expensive diesel engines (EPA 2004b). Such behavioral aspects were only discussed qualitatively in the report. Incorporating behavior into environmental regulatory models is discussed more generally in Box 2-1.

Linkages among the different processes are not seamless. Each category often is represented by a separate model and regulatory analyses often require that inputs and outputs of one model interface with other models in separate categories. Sometimes temporal or spatial scales do not line up and results from one model may not have natural counterparts in the models with which it interfaces. An example is from the air quality analysis in which emissions from vehicles and other sources that are estimated at the regional level must be allocated spatially and estimates of aggregated hydrocarbon emissions must be disaggregated by species for input into the air quality model. More fundamentally, the linking of these different categories means the linking of separate disciplines. To properly link different modeling categories requires the building of interdisciplinary bridges, which is an ongoing effort at EPA. Although there are software tools and integrated models that allow multiple processes to be combined into a single modeling framework as discussed in a subsequent section, such a model still faces the difficulty of needing to rely on the expertise from multiple disciplines.

The level of effort dedicated to environmental regulatory applications varies greatly. This variation is a critical consideration when

BOX 2-1 Incorporating Human Behavior into Environmental Models

For regulatory purposes it is important to not only model natural systems but also human activities and their interactions with natural systems. These interactions, which can always be found at either end of the causal chain, shown in Figure 2-1, and often in the middle as well, require models from the social sciences, usually economics. A key modeling consideration is the extent to which such models incorporate human behavior. The earliest models used for environmental regulatory purposes had little if any behavioral content. The effects of both regulations and environmental changes were estimated without considering the full range of responses available to economic agents—individuals, households, and firms. One of the first models to demonstrate that possible behavioral responses could affect the costs or effectiveness of regulations was developed by Gruenspecht (1982), who pointed out that the common regulatory practice of requiring more stringent and more costly abatement for new sources of pollutants than for existing sources could retard the turnover of existing equipment. Behavioral responses are sensitive to the details of regulatory design, and numerous models appeared in the economics literature describing the unintended consequences of such real-world policies as CAFE (Kwoka 1983) and vehicle inspection and maintenance (Hubbard 1998). Behavioral responses also affect other outcomes of interest to EPA, including regulatory enforcement (Harrington 1989), pollution abatement subsidies (Freeman 1978; Rubin 1985). Behavioral responses to adverse environmental consequences, such as private defensive expenditures, have also been analyzed.

For many years, EPA made frequent use of behavioral models for policy analysis and regulatory impact analysis. In cases involving economic incentives, behavioral models are essential because the behavioral response is what drives the policy outcome. For example, analysis of proposed emissions cap-and-trade policies to control airborne sulfur dioxide emissions from the electric power industry requires the agency to predict the behavior of utilities in the permit market. For this task, EPA uses the integrated planning model, a proprietary dynamic linear programming model that determines the least-cost loading of generating capacity to meet electricity demand. The optimization simulates the expected outcome in the permit market.

Not all of EPA's regulatory models that could incorporate behavioral responses to regulation do. For example, the MOBILE model, which projects average regional or national motor vehicle-emission rates under a variety of regulatory design parameters, does not consider the effects that regulatory alternatives might have on fleet composition or vehicle use through their effects on vehicle or fuel prices. MOBILE's failure to anticipate behavioral responses to regulation has been most noticeable in the motor vehicle emissions inspection and maintenance program (I/M) component, which has underestimated the ability of motorists to avoid I/M tests altogether and overestimated the ability of those tests to identify high-emitting vehicles as well as the effectiveness of vehicle repair (e.g., NRC 2001a; Holmes and Cicerone 2002, and references therein).

developing recommendations, later in the report, related to model development, evaluation, and application. At one end of the spectrum are applications that involve a small investment in resources and modeling effort. Leaking underground petroleum storage tanks number in the hundreds of thousands, and preliminary screening for EPA's leaking underground storage tank program typically relies on the application of an analytical model with assumed parameters (Weaver 2004). These state-run programs may spend as little as $500 for site assessments. The new chemicals program under the Toxic Substances Control Act (TSCA) requires EPA to review approximately 2,000 new chemicals per year and issue decisions on up to 20-30 chemicals per day (C. Fehrenbacher personal commun., EPA Office of Pollution Prevention and Toxics, February 23, 2006). Because of these demands, the agency relies on the quantitative structure-activity relationships (QSARs) model that uses basic knowledge of a chemical's structure to predict physical and chemical properties and environmental fate and transport when data are not available. At the other end of the spectrum, EPA may spend years or even a decade assessing the health and environmental consequences of other environmental pollutants, making their modeling efforts extremely complex. Under the CAA, EPA is required to review NAAQS every 5 years. This requires major investments of resources and may take many years of assembling background information and performing analyses, including modeling analyses. Somewhere between these two extremes are the water quality management TMDL and the air quality management SIP analyses. EPA estimates 3,000-4,000 TMDLs, with a wide array of resource requirements, will be needed annually for the next 8 to 13 years to meet current deadlines (NRC 2001c). While some TMDLs require extensive data collection and modeling, at least one state has proposed using a nonmodeling approach for catchments with little or no data (George 2004). The SIP process can be a major undertaking requiring development of emissions inventories and analysis of control options. Each local area out of attainment must submit a plan for each pollutant. For example, there are currently 116 counties out of attainment with the current 24-hour $PM_{2.5}$[2] standard (Bachman 2006).

---

[2]$PM_{2.5}$ refers to a subset of particulate matter collected by a sampling device with a size-selective inlet that has a 50% collection efficiency for particles with an aerodynamic diameter of 2.5 μm.

Evolution of Regulatory Modeling

The elements that are included in modeling may change over time for a given type of assessment, typically adding complexity to the modeling process. This is a result of changes to regulatory requirements, scientific understandings, and modeling capabilities. A potential example is in the health risk assessment paradigm. Fundamentally, a health risk assessment developed today is conceptually consistent with what is discussed in the NRC "Red Book" (*Risk Assessment in the Federal Government: Managing the Process*, NRC 1983) and laid out in Figure 2-2. A major modeling component is the development of dose-response relationships through analysis of epidemiological or toxicological studies (Setzer 2005). The NRC reports on toxicological effects of arsenic from drinking water provide a prime example of many of the issues associated with developing dose-response modeling for a contaminant (NRC 1999a, 2001b). Box 2-2 described this case study in more detail.

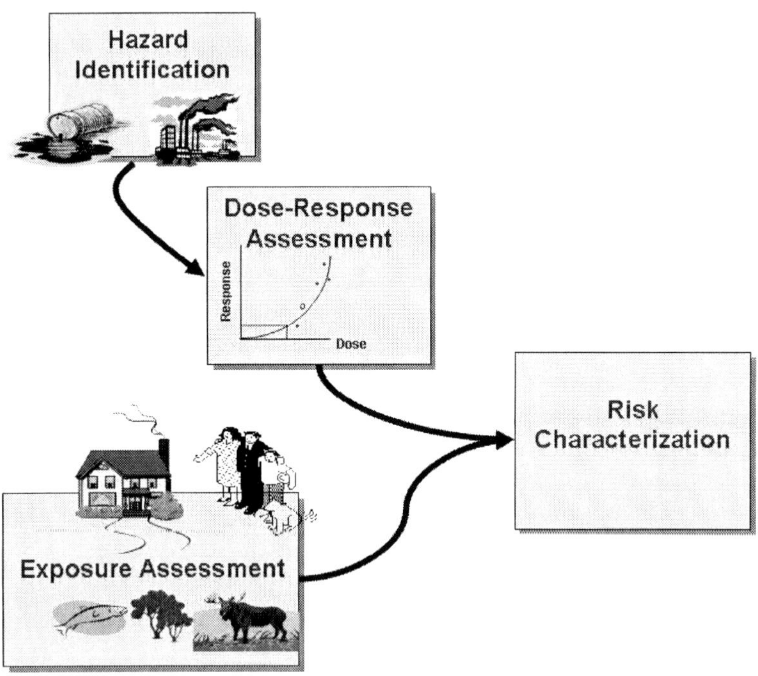

FIGURE 2-2 Basic elements of risk assessment from the National Research Council's Red Book. Source: NRC 1983.

BOX 2-2 Risk Assessment for Arsenic in Drinking Water

EPA's 2001 risk assessment for arsenic in drinking water provides a rich case study that illustrates many of the challenges associated with using models to inform environmental regulation. Establishing a U.S. standard for arsenic in drinking water has been a source of controversy for many decades. From the perspective of environmental regulation, the arsenic story is an interesting one for a number of reasons. First, exposures arise from natural sources and some have even argued that at very low doses it is an essential element for human health. Second, arsenic is not directly carcinogenic in animals; hence, all evidence for human health effects arises from epidemiological studies.

Two National Research Council committees (NRC 1999a, 2001b) convened to advise EPA on this matter suggested that regulation be based on data from 42 villages in southwestern Taiwan, which showed increased rates of bladder and lung cancer as a function of arsenic levels measured in village wells. While it was originally hoped that the arsenic might provide an opportunity for using EPA's then new guidance on carcinogen risk assessment that allowed the use of biologically based models, the first NRC committee found that there was so much controversy over underlying mechanisms that it was not possible to identify a suitable biologically based model. Instead, the committee recommended reliance on more empirically based statistical models. Although the dose-response modeling was based on human data, which removed the uncertainty associated with extrapolation of results from animals to humans, the inherent variability associated with human data introduced other sources of uncertainty. There were many concerns expressed about the appropriateness of relying on the Taiwanese data for the purpose of setting regulations in the U.S. context. Some cited differences in dietary patterns between the United States and Taiwan, particularly in this relatively poor rural area of Taiwan. Others were concerned that the Taiwanese study used cancer incidence data extracted from population records and exposure crudely assessed based on the median levels of arsenic measured in villages wells. Indeed, Morales et al. (2000) fit the data using a series of relatively simple empirical models that differed according to how age and exposure were incorporated and compared the results obtained from the multistage Weibull model, which had been classically suggested for the analysis of time-to-event data of the type encountered in the Taiwanese data set. As shown in Figure 1-5, these various models differed substantially in their fitted values, especially in the critical low-dose area that is so important for establishing the benchmark dose used to set a reference dose (RfD).

Rule-making was able to move forward, despite the uncertainty, since all the models supported the conclusion that risk levels at the then standard of 50 micrograms per liter (μg/L) were unacceptably high. Based on the first NRC review, EPA lowered the standard for drinking water from 50 μg/L to 10 μg/L in January 2001. This standard was initially delayed so that the EPA Science Advisory Board, the National Drinking Water Advisory Council, and a second NRC committee could further examine benefits, costs, and health risks. These reviews supported the proposed 10 μg/L standard, which was subsequently finalized by EPA.

However, research and practice has enabled major changes since 1983 in how the risk components are developed. As the black box between exposure and effect gathered light, improvements in risk assessment practice (Reddy et al. 2005), toxicological testing technologies (NRC 2006b), biomonitoring (NRC 2006a), and understanding of the modes of action, to name a few allow for a more mechanistic modeling approach to relating exposures to health outcomes (see Figure 2-3). For example, analysts understand that even if humans and rats received the same external exposure, they did not receive the same dose of active chemical to the target tissue. To understand these events, data on basic physiological and pharmacokinetic processes and resultant physiologically based pharmacokinetic (PBPK) modeling can integrate specific properties of chemicals with age- and organ-specific physiological processes in different species to relate an effective experimental exposure to an effective environmental exposure (Clewell 2005). PBPK models offer the possibility that doses of chemicals delivered to target cells of a rat could be quantitatively extrapolated to target cells of humans. These models also offer the possibility that differences in age or sensitivity (for example, polymorphisms in metabolism) in the human population could be incorporated in models. However, the use of a PBPK model in a risk assessment can be a time- and cost-intensive undertaking requiring expertise (Clewell 2005). It must be accompanied by a thorough evaluation that includes the following:

- Evaluation of biological plausibility of model structure and parameters.
- Verification of model code (equations and logic).
- Validation of model's region of applicability.
- Sensitivity and uncertainty analysis.

These models may also bring with them many technical and science policy challenges. One outcome of more mechanistic approaches to health risk assessment modeling is that regulatory end points might be based on an upstream biochemical precursor event instead of observed adverse health outcomes. The challenge (and controversy) then becomes selecting the appropriate point between an innocuous molecular change and frank disease to use in the assessment. The risk assessment of perchlorate (NRC 2005b) offers an example of how this can be addressed by selecting a nonadverse effect (the inhibition of iodide uptake by the

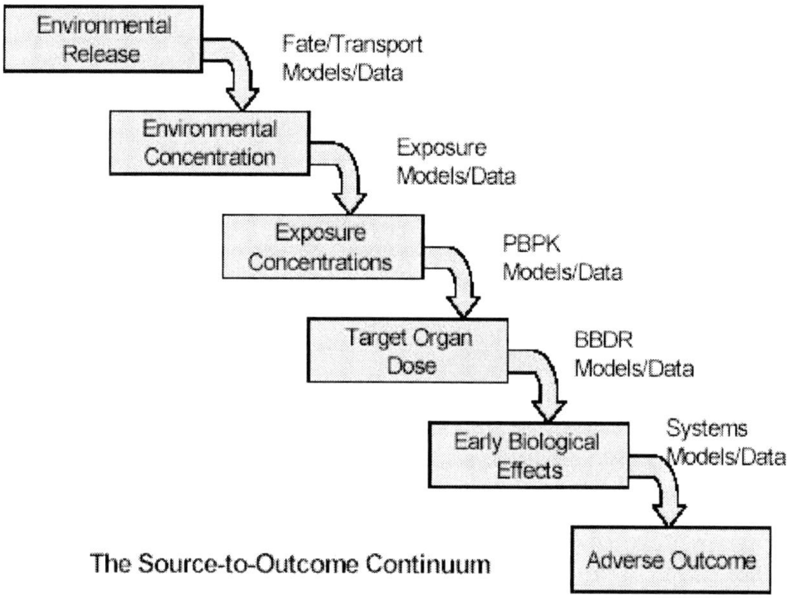

FIGURE 2-3 Elements of advanced mechanistic approaches to health risk assessment. This figure illustrates the fundamental elements of assessment and the models that link the elements. PBPK refers to physiologically based pharmacokinetic models and BBDR refers to biologically based dose-response models. Source: EPA 2003b.

thyroid gland) as a point of departure for adding uncertainty factors. More fundamentally, as understanding increases, so do options and questions about the most appropriate approaches to assess risks. For example, in a particular scenario, judgments may be needed as to whether EPA should give preference to empirical models using human epidemiology or mechanistic rodent-based models (Preuss 2006).

Integrated Models and Modeling Frameworks

Some models tend to fit into a single category, while other regulatory models represent multiple categories of processes, such as modeling emissions and fate and transport together. For example, the integrated planning model produces estimates of electricity sector activity, including fuel demands, prices, and emission-control decisions for given levels of emissions (Napolitano and Lieberman 2004). Models that represent pesticide exposure, such as the CARES, DEEM/Calendex, and LifeLine models, simulate activities that expose humans to pesticides and the residues that different pesticides produce in food and the residential environment to simulate exposure profiles (EPA 2004d).

More recently, the movement toward integration has utilized advances in software to develop modeling frameworks that allow user flexibility to use a combination of compatible models, facilitate multiple simulations, and facilitate output analysis. Examples are CMAQ/Models-3, FRAMES, 3MRA, and BASINS. For example, the Better Assessment Science Integrating Point and Nonpoint Sources (BASINS) is a multipurpose environmental analysis system that integrates a geographical information system, national watershed data, and state-of-the-science environmental assessment and modeling tools into one modeling package (EPA 2006e). The model integrates individual stand-alone models that simulate pollutant loadings from point and nonpoint sources and in-stream water quality models for performing watershed- and water-quality-based studies. It is intended to make watershed and water quality studies easier by bringing key data and analytical components "under one roof." A further discussion of improvements in integrated model methods is contained in Chapter 6.

Regulatory Models from a Regulatory Perspective

In this section, we describe the use of models in six phases of the regulatory process. Strategic planning identifies environmental problems of present and future importance and assembles data and constructs modeling tools to permit analysis. Rule-making translates congressional directives into specific regulations. Delegation has states and localities given responsibilities for developing plans to achieve environmental goals locally and writing regulations to achieve those goals. Permitting, licensing, registration is where these rules are applied to govern the behavior of polluting individuals, firms, or other entities. The last two phases are enforcement and ex post facto analysis.

Strategic Planning

The first element in the regulatory sequence above involves the strategic use of models to inform Congress and decision makers within

EPA in deciding whether or how to legislate or regulate. "Strategic" implies a thoughtful, informed, and priority-set analysis that identifies goals and major approaches to achieve the goals. Because strategies are inherently predictive, models are crucial. They can inform the identification of goals that are important to achieve (for example, whether a certain air pollutant already regulated is still an important public health risk requiring additional legislation or regulations), and they can characterize approaches to achieving them (for example, whether the predominant source of this air pollutant is stationary, mobile, or personal identifies optimal regulatory targets). Examples include congressional requests to assess alternative legislative proposals for controlling multiple pollutants from power plants (EPA 2001a,b) and EPA's internal use of modeling to identify the population at risk from ozone exposure that guided decisions on changing the NAAQS for this pollutant (EPA 1996). The use of modeling in strategic planning can become part of the debate between Congress and EPA over environmental policy. An example of this is a May 13, 2004, letter from Congressman Thomas Allen to EPA Administrator Michael Leavitt concerning delays of model runs assessing control options for electric power-plant emissions of mercury (Allen 2004).

One of the broadest uses of models in environmental strategic planning was by the congressionally mandated National Acid Precipitation Assessment Program (NAPAP), which was directed to perform research to inform decisions on regulations of acid rain. The interagency program (EPA and 11 other federal agencies) was funded for 10 years in the 1980s and produced 27 state-of-the-science and -technology reports on all aspects of the acid rain issue. One of the primary products was the air quality models that are precursors to the models used at EPA today. Information developed by NAPAP was useful for changing the understanding of the scientific-information related to acid rain and informing Congress in its development of the parts of the CAA Amendments of 1990 that dealt with acid rain. However, this legislation was enacted before NAPAP completed its integrated assessment report of its activities. This report was intended to synthesize the science for policy makers (NAPAP 1990). In the end, NAPAP was criticized on a number of different levels by both the participants and the observers (Roberts 1991; Rubin 1992; Herrick 2000). Global warming modeling provides a contemporary example of the strategic use of models. The U.S. Climate Change Science Program (CCSP) has developed a strategic plan for attempting to coordinate research, including modeling research, being done by 13 agencies and departments in the government (CCSP 2003; NRC 2004b).

Rule-making

Rule-making encompasses the tasks of regulatory design and promulgation. The goal of regulatory design is to produce a proposed rule that complies with the legislative requirements set down by Congress and that provides sufficient support and analysis of the rule. EPA's modeling activities at the rule-making stage can be extensive. For example, the nonroad diesel RIA included the use of activity models, emissions models, air quality models, engineering cost models, energy forecasting models, petroleum refinery models, and human health and agricultural impacts models to assess the benefits and costs of the proposed regulation (EPA 2004b). Other rules incorporate less modeling. However, at this point in the regulatory process, EPA is responsible for performing the model analysis, although other stakeholders may submit model analysis and comments on the agency's modeling analysis during the public comment period. The external review of EPA's modeling in support of rule-making, including the role of the public comment period and interagency review, is discussed in a later section of this report.

Delegation

Many environmental statutes, including the CAA and CWA, delegate important roles for compliance, which includes implementation and enforcement, to states. States may further delegate some responsibilities to local agencies. Delegation of authority for implementation and enforcement is also given to tribal governments. Modeling analysis is part of the delegated responsibility. The roles of EPA and the state and local agencies vary by the statutes and within statutes. Under the SIP process, states or local governments must prepare a plan for each area that does not meet NAAQS, describing how that area will be brought into attainment. This process includes the modeling analysis described in Table 2-2. For large urban areas, typically a metropolitan planning or other local air quality agency prepares the SIP that must be then approved by the state and eventually by EPA. In the case of the Los Angeles area, it is the South Coast Air Quality Management District that prepares the SIP. For areas with smaller populations, such as the Missoula, Montana, area, the state prepares any SIPs submitted for approval to EPA. For the TMDL program, states are primarily responsible for carrying out the program, including the modeling described in Table 2-2. However, EPA will carry out some TMDL for particularly contentious settings, such as the establishment of a TMDL for limiting mercury in fish tissue residual in the Ochlockonee River Watershed in Georgia (EPA 2002b). Thirty-five states run their own programs for dealing with leaking underground storage tanks, including assessments of subsurface containment transport and risk assessment modeling. As mentioned, tribal governments have the option of running their own environmental programs, and some tribes have received authorization to run air quality and water quality programs. Private consultants often are engaged to perform part of the modeling analysis required under state delegated programs.

State-generated source-specific regulations, required by both SIP and TMDL, are based on the effects of air and water pollutants on environmental quality. This requirement raises a host of technical, economic, and political issues that are sometimes not sufficiently covered in the writing of federal standards. The issues include the following:

- *Interdependence.* The environmental effects of emissions from any one source depend on the emissions from numerous other sources.
- *Nonpoint sources.* Emissions from sources that are difficult to monitor and regulate at the individual level either because the sources are numerous and diffuse or because the emissions are episodic and dependent on natural processes.
- *Distributional asymmetry.* The sources responsible for pollutant discharges are located in a different area from where the environmental damages are suffered. For example, states and cities may have no control over air pollutants that have blown in from afar.

Permitting

Other statutes, such as the Toxic Substances Control Act, Safe Drinking Water Act, and Food Quality Protection Act, require EPA or the state to permit an activity. This activity might be required for the construction and operation of a point emissions source or the introduction and continued use of a chemical in the market. The statues vary in what role modeling plays and which entities perform the modeling. For licensing new pesticides, manufacturers supply a substantial amount of modeling of environmental and human health risks to EPA that might be supplemented by additional agency analysis. For the relicensing of pesticides, which is carried out under the Food Quality Protection Act's mandate to assess cumulative and aggregate risks, EPA performs the modeling analysis. For the premanufacturing determination that must be made before new chemicals can enter the market under the Toxic Substances Control Act, EPA is responsible for assessing risks. The initial screening is done using structure-activity models, and the results of such modeling determine whether a more thorough assessment is needed and whether manufacturers will be required to submit more test data. Programs that permit discharges into water, controlled under the National Pollution Discharge Elimination System (NPDES), are primarily run by the states, although some states have only partial authority. Although many of the requirements under the NPDES program are still driven by technology-based standards, increasingly state and federal permit writers must take into account water quality standards and watershed considerations, which increases modeling needs. The CAA mandates that the states implement and that EPA oversee permit programs to control and regulate pollutant emissions from major stationary sources. Under these programs, each new major stationary source of air pollutants must apply for a permit before construction and provide modeling to help to demonstrate that the new facility will meet appropriate emission-control standards. Permittee modeling is subsequently reviewed by state regulators.

Compliance and Enforcement

Models are used in compliance and enforcement in several ways. For enforcing some regulations, EPA uses models to estimate the benefit to the regulated party—usually cost savings—from delaying or avoiding pollution control expenditures. For example, the BEN model (see EPA 2007) calculates a violator's economic savings from delaying or avoiding pollution control expenditures. This estimate is then used as a basis for setting the penalty, which ensures that the violation will not be to the regulated party's advantage. Other models assess a regulated party's ability to afford such costs as civil penalties, Superfund cleanup costs, and pollution control expenditures. An example is the MUNIPAY model (EPA 2006f), which evaluates a municipality's or regional utility's ability to afford compliance costs, cleanup costs, or civil penalties. EPA may also use models to estimate "natural resource damages" from private actions that damage natural resources. These natural resource damage actions arise out of legislative liability schemes under the CWA, the Oil Pollution Act, and the Comprehensive Environmental Response, Compensation and Liability Act. The damage estimates are generally based on contingent valuation surveys, as well as models that attempt to estimate the costs of restoring or replacing the damaged resources.

Ex Post Facto Auditing and Accounting of Impacts

Like strategic planning, assessment of the performance and costs of regulations after they have been implemented is relatively rare within EPA, although it is often carried out by other parties. The Office of Management and Budget (OMB 2005) reviewed recent ex post facto analyses of regulations, including environmental regulations. EPA has also received periodic requests from Congress to report on the aggregate costs and benefits of its regulations. In the past, for example, Congress has required EPA to periodically estimate the total costs of the CAA (under section 812) and CWA (under section 512). One of the main sources of data for these reports is the Pollution Abatement and Control Expenditure (PACE) Survey conducted annually from 1978 to 1994 and conducted in 1999 by the U.S. Census Bureau on a sample of manufacturing establishments.[3] In the PACE Survey, individual establishments report total expenditures on pollution abatement, separated by receiving medium (air, water, or land). EPA uses the survey results to estimate expenditures for all manufacturing plants and adds information on expenditures in other sectors to produce the report. These analyses are of limited use for policy assessment because they report only on the aggregate costs of regulation rather than the costs of specific regulations.

Similarly, EPA also occasionally conducts ex post facto studies of benefits. The most prominent example is the ongoing study of the benefits and costs of the CAA—a study required by section 812 of the 1990 CAA. The first major report was a retrospective study of the benefits and costs of the CAA from 1970 to 1990 (EPA 1997a). This report was followed by a prospective study of the benefits and costs of the CAA from 1990 to 2010 (EPA 1999a). A second prospective study is in progress for the period from 2000 to 2020. The retrospective study is best known for its controversial benefit estimate of $6 to $50 trillion in benefits over the period. It illustrates the difficulty of estimating benefits and costs of massive, aggregate programs such as the CAA. All benefit and cost estimates require comparison to a "without-regulation" scenario. For very large changes, determining an appropriate without-regulation scenario be-

[3]A new PACE survey is in development at the EPA and may soon resume.

comes a matter of achieving consensus rather than analysis and is riddled with uncertainty. Modelers outside of federal agencies also contribute post hoc analysis of environmental regulatory activities. The literature is vast. Some particular examples include the assessments of compliance costs and other impacts of the sulfur dioxide emissions trading programs (e.g., Ellerman et al. 1997; Stavins 1998; Burtraw and Mansur 1999) and the effects of "corporate average fuel efficiency" standards on energy consumption and emissions from motor vehicles (e.g., Kliet 1990; Harrington 1997; Greene et al. 1999; Portney et al. 2003).

CONGRESSIONAL AND EXECUTIVE BRANCH INFLUENCES

There are some particular influences and constraints on the regulatory process resulting from the enabling statutes passed by Congress and from a series of executive orders that over time have given OMB oversight responsibility over regulations and imposed specific requirements on how regulatory decisions are supported through modeling. It is essential to understand these influences.

Congressional Influence

Federal environmental statutes, such as the CAA and CWA, usually contain statements of health and welfare goals, schedules and deadlines for meeting them, and, often, criteria for determining whether the goal has been met. Table 2-3 contains a sample of some of the general and specific directives found in several important environmental statutes. To write regulations to meet these requirements, EPA produces much analysis to justify its decisions and show how its actions meet the congressional directives, which can sometimes require the agency to do the following:

- Explain quantitatively the magnitudes as well as the spatial and temporal patterns of present and projected contamination.
- Trace the contaminant back to the human activities that contribute to the contamination and trace the contaminant forward to its health impacts.

TABLE 2-3 Examples of Substantive Legislative Directions for EPA Models

General Directions	
Toxic Substances Control Act, 15 U.S.C. § 2605(a)	Authorizing regulatory action on existing toxic substances "if the administrator finds that there is a reasonable basis to conclude that the manufacture, processing, distribution in commerce, use, or disposal of a chemical substance or mixture, or that any combination of such activities presents or will present an unreasonable risk of injury to health or the environment").
Clean Air Act, 42 U.S.C. § 7409(b)(1)	NAAQS for criteria pollutants must "protect the public health," "allowing an adequate margin of safety."
Federal Insecticide, Fungicide, and Rodenticide Act, 7 U.S.C. § 136a(c)(5)(D)	Allows pesticides to be registered only if the administrator finds that "when used in accordance with widespread and commonly recognized practice it will not generally cause unreasonable adverse effects on the environment."
Federal Food, Drug, and Cosmetic Act, 21 U.S.C. § 346a	A protective standard for pesticide residues is rebutted only once "there is a reasonable certainty that no harm will result from aggregate exposure to these residues."
Safe Drinking Water Act, 42 U.S.C. § 1412(b)(4)	Maximum drinking-water contaminants are "set at the level at which no known or anticipated adverse effects on the health of persons occur and which allows an adequate margin of safety."
Clean Water Act, 33 U.S.C. § 1313(c)(2)(A)	The objective of the Act is to "restore and maintain the chemical, physical, and biological integrity of our Nation's waters." Water quality standards set by statute "shall be such as to protect the public health or welfare…."
Resource Conservation and Recovery Act, 42 U.S.C. § 6924(m)	Standards for treatment of hazardous wastes disposed onto land shall specify "those levels or methods of treatment, if any, which substantially diminish the toxicity of the waste or substantially reduce the likelihood of migration of hazardous constituents from the waste so that short-term and long-term threats to human health and the environment are minimized."

(Continued)

TABLE 2-3 Continued

General Directions	
Comprehensive Environmental Response, Compensation, and Liability Act, 42 U.S.C. § 9621(b)	"The President shall select a remedial action that is protective of human health and the environment, that is cost effective, and that utilizes permanent solutions and alternative treatment technologies or resource recovery technologies to the maximum extent practicable" and specifying additional criteria the President must consider in selecting a remedial action.
Specific Directions	
Food Quality Protection Act of 1996, 21 U.S.C. 346a(b)(2)(C) and (D)	"In the case of threshold effects … an additional ten-fold margin of safety for the pesticide chemical residue shall be applied for infants and children" … with additional legislative specifications for the types of information that must be used in conducting the risk assessment.
Safe Drinking Water Act, 42 U.S.C. § 300g-1 (b)(3)(B)	"The Administrator shall, in a document made available to the public in support of a regulation promulgated under this section, specify, to the extent practicable: i) each population addressed by any estimate of public health effects; ii) the expected risk or central estimate of risk for the specific populations; iii) each appropriate upper-bound or lower-bound estimate of risk…."
Resource Conservation and Recovery Act, 42 U.S.C. § 6924(g)(10).	Requiring (for example) the Administrator to "complete a study of hazardous waste managed [with specific types of treatment processes] … to characterize the risks to human health or the environment associated with such management … [n]ot later than five years after March 26, 1996."

Source: EPA 2004a.

- Project patterns of contamination and their impacts under various regulatory proposals (including no regulation and, in some cases, deregulation).

To produce the kind of regulations authorized by such health- or welfare-oriented legislation, therefore, requires the use of the types of models discussed in the preceding section and displayed in Figure 2-3. The figure, to be sure, suggests a degree of simplicity that EPA does not necessarily enjoy in its regulatory activities. EPA must translate general and sometimes vague statutory prescriptions into specific rules governing the behavior of individuals, firms, and state and local governments; pollutant sources must be identified and brought into compliance with the rules; and periodic assessments must be undertaken to ensure that satisfactory progress is being made to meet the statutory goals. Notions like "restore and maintain the chemical, physical, and biological integrity of our Nation's waters" must be put into regulatory practice. Such legislative mandates often require EPA to develop or use models despite substantial data gaps and minimal supporting theory. For example, besides requiring the use of MACT standards for HAPs, the 1990 CAA Amendments also required a secondary regulatory phase when EPA is instructed to assess the "residual risk" due to a HAP that remains after compliance with the standards. Besides the need to interpret the meaning of the term "residual risk," there are many technical difficulties associated with assessing such a risk, including the methods of calculating risks and data limitations (NRC 2004a). A similar modeling challenge occurs with the mandates in the Food Quality Protection Act that requires EPA to assess aggregate health risks from exposure to one chemical from multiple pathways and cumulative health risks from aggregate exposure to multiple pesticides (EPA 2001c, 2002d). Though it may pose difficulties for modelers, the agency's priority must be to ensure that its regulations meet the requirements set forth under the legislation, not whether the regulations fit model capabilities.

Legislation also affects how EPA uses model assumptions. For example, under the CAA, EPA is instructed to set NAAQS for criteria pollutants that are "requisite to protect the public health" with an "adequate margin of safety."[4] This mandate has been interpreted by the

---

[4]Criteria pollutants are air pollutants emitted from numerous or diverse stationary or mobile sources for which NAAQS have been set to protect human health and public welfare. The criteria pollutants are ozone, particulate matter, carbon monoxide, sulfur dioxide, nitrogen dioxide, and lead.

courts to require EPA to use models in such a way that their results are more likely to err on the side of caution with respect to protecting public health and to prohibit the agency from taking economic costs into account in setting standards.[5] The adequate margin of safety is a policy choice of the EPA administrator's intended to protect sensitive groups from adverse health effects (Murphy and Richmond 2004). The impact on regulatory modeling is that control costs, technological feasibility, and cost-benefit comparisons are not included in the analysis used to set NAAQS. It also causes EPA to consider a variety of sources of modeling evidence as shown in Figure 2-4.

Executive Branch Oversight

While being overseen by the White House, Congress, and others, EPA exercises substantial discretionary authority to implement and enforce environmental laws. With respect to models, EPA makes the vast majority of decisions about whether a model is needed to implement or enforce a legislative mandate, how to select and review models to carry out its authorities, and when it is time to replace one model with another.

The executive branch has provided oversight of the regulatory process through analytical requirements for the review of the costs, benefits, and effects of all major regulations. This factor has produced extensive modeling requirements for major regulatory actions overseen primarily by the Office of Information and Regulatory Affairs within OMB. One requirement is for an assessment of benefits and costs for major regulations through an RIA. Box 2-3 discusses the history of the RIA requirement.

For an RIA to be required, a regulation must have estimated economic effects that exceed $100 million annually or must have important adverse effects on prices, employment, productivity, or other economic consequences. Few regulations issued by EPA or other agencies require an RIA; in FY2004, for example, 4,088 rules were published in the *Federal Register*, but only 11 had RIAs. Of those 11, 6 were issued by EPA, 4 were issued by the Office of Air and Radiation, and 2 were issued by the Office of Water. Despite the small numbers, OMB estimates that the rules requiring RIAs "capture the vast majority of total costs and benefits

---

[5] *Lead Indus. Ass'n v. EPA*, 647 F.2d 1148 (D.C. Cir. 1980); *American Trucking Ass'ns v. EPA*, 175 F.3d 1027, 1034 (D.C. Cir. 1999).

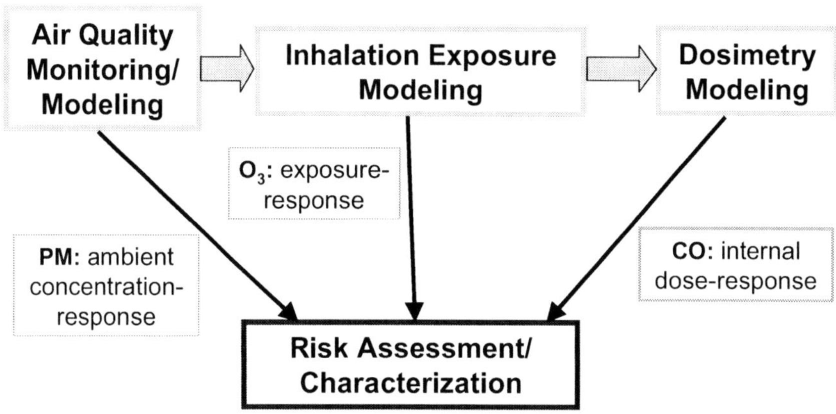

FIGURE 2-4 Sources of information for setting various NAAQS. Source: Murphy and Richmond 2004.

BOX 2-3 The Development of the Requirement for Regulatory Impact Analysis for Major Federal Rules

RIAs are required currently for any regulation whose estimated economic effects (costs) exceed $100 million annually or have important adverse effects on prices, employment, productivity, or other economic consequences. The requirement for an RIA came about as Presidents sought to have more influence over the agendas of executive agencies by requiring a review of the costs, benefits, and effects of all major regulations. The key event was Executive Order 12291 (EO12291), issued on February 17, 1981, announcing new rules governing the issuance of regulations by federal agencies. EO12291 introduced two important innovations into federal rule-making. First, it required federal agencies to produce, before certain "major" proposed regulation could appear in the *Federal Register*, an assessment of the benefits and costs of the proposal and alternatives to it. Before this executive order, economic assessment of regulations was concerned not with benefits and costs but with "economic impacts," which included the effect of the regulation on inflation, employment, and the profits of affected industries.[6] In addition, EO12291 required centralized review of regulations and the accompanying RIA by an oversight group, the Office of Information and Regulatory Affairs (OIRA) housed in OMB.

Each President since has either issued his own executive order affirming the RIA requirement and the OMB review or accepted that of his predecessor. For example, EO12866, issued on September 30, 1993, changed the procedure to increase the public's accessibility, added requirements to specifically address

(Continued on next page)

[6]See Magat et al. (1986) for a discussion of the preparation and use of such studies in the Effluent Guidelines rule-making process.

the problem to be addressed by the regulation (usually a market failure) and examine distributional consequences of new rules, and require only that the benefits of proposed regulations have to "justify" the costs, not "outweigh" the costs as it had been in EO12291. For the most part, recent Presidents of both parties have retained support for regulatory review requirements, including the RIA.

The implication of the RIAs for EPA modeling is that where possible, all the effects of a proposed regulation, positive and negative, must be expressed in monetary terms. Since most of the benefits—and many of the costs—of environmental regulation are not traded in markets, econometric models are needed to estimate individuals' willingness to pay for the predicted physical effects of regulations, such as improved air quality.[7] The RIAs could result in the estimation of regulatory benefits and costs even for rules where the enabling legislation has expressly forbidden the use of costs to make regulatory decisions. For example, as noted in the preceding section, the CAA prohibits cost to be a criterion in the setting of NAAQS. However, that did not prevent a very extensive and thorough RIA from being prepared to support the 1997 revision of the ambient standards for ozone and fine particulates (EPA 2006g). The RIA found very large positive net benefits for both standards, so there was no actual conflict between the RIA requirement that the costs be justified by the benefits and the legislative conflict that costs not be considered. The most recent OMB guidance on the preparation of RIAs is in OMB Circular A-4 (OMB 2003), which has expanded the requirements for uncertainty analysis.

of all rules subject to OMB review" (OMB 2005). In addition, rules exceeding $1 billion per year in economic effects are subject to a further requirement to include a formal analysis of uncertainty. Only the nonroad diesel rule in FY 2004 was subject to this requirement (OMB 2005). As discussed in Chapter 4, uncertainty analysis adds considerably to the analytical burden of producing and comparing alternative regulations with unclear benefits.

In addition, the executive branch has been interested in the quality of information and peer review practices used by federal agencies, including EPA. One set of guidelines developed by OMB is *Guidelines for Ensuring and Maximizing the Quality, Objectivity, Utility, and Integrity of Information Disseminated by Federal Agencies* (OMB 2001). These guidelines, which were mandated by the Information Quality Act (IQA) (Treasury and General Government Appropriations Act for Fiscal Year 2001, Pub. L. No. 106-554, § 515, 114 Stat. 2763 [2000]), called for

[7]A snapshot of the state of the art in valuing mortality and morbidity reductions (by far the most important source of monetizable environmental benefits) can be found in the proceedings of a workshop sponsored by EPA's National Center for Environmental Economics (EPA 2006h).

agencies to issue information-quality guidelines to ensure the quality, objectivity, utility, and integrity of information. Recognizing the critical roles that models have in developing information, EPA has developed its own guidelines for data use to ensure that the models used in regulatory proceedings are objective, transparent, and reproducible (EPA 2002a). In addition, as discussed in a later section, OMB has released guidance on peer review (OMB 2004).

OVERSIGHT PROCESSES GOVERNING REGULATORY MODELS AT EPA

After Congress or EPA has decided to use a model for one or more regulation-relevant purposes, the model normally goes through some internal and external oversight to ensure that it meets scientific, stakeholder, and public approval. Although these oversight processes are not perfect and run the risk of introducing their own sources of error or complication, they nevertheless exert an important and independent pressure on regulatory models that is generally not present when models are developed and used in nonregulatory settings.

Because the results of models can impose important costs on regulated parties and the public at large, EPA's evaluation of models used for regulatory design and promulgation (the rule-making phase from above) is the most heavily constrained by legislative requirements, regulatory review, and legal challenges. Figure 2-5 shows a schematic of the regulatory requirements placed on the regulatory design and promulgation phase. Models used for other regulatory purposes—outside of rule-makings—are generally not subjected to these extensive internal and external review requirements. Models used at the enforcement stage, for example, are generally not required to go through peer review or even notice and comment, but they are required to at least gain judicial acceptance before a court will enter penalties against a violator based on the model. Models used in environmental regulatory programs delegated to the states, such as models used to develop SIPs and TMDLs, can be subjected to public comments and debate, but independent peer reviews of individual model applications are not required. Models used in providing guidance may be subjected to scientific and public review, but generally, this review is done at the agency's discretion. The Science Advisory Board and Science Advisory Panel, described in a subsequent section, are two sources for peer reviews. Models used for strategic

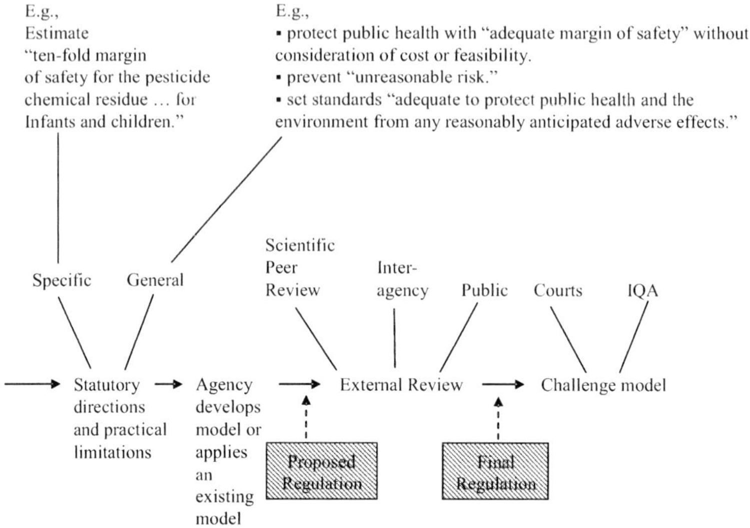

FIGURE 2-5 Flow chart of general regulatory requirements for models used at the regulatory design and promulgation stage.

planning or priority setting within the agency involve even fewer mandatory oversight processes. Yet in these cases, EPA still develops guidelines for internal peer review and may voluntarily subject these models to sources of external review as well.

Because regulatory design models encounter the most extensive oversight requirements and also tend to be an important modeling activity at EPA, regulatory design models are the focus of the remaining discussion. In general, these models require multiple layers of review, including formal scientific peer review, notice and comment processes, and intra-agency review. Interested parties are also provided with an opportunity to challenge the model to the agency and in court to ensure that the model is reliable.

External Review of EPA's Models

The first and perhaps most important set of requirements involves subjecting regulatory decisions, including the models underlying them, to review by three layers of outside reviewers. This external review is thus conducted independently of the authors of the model or the users for a specific application. This section summarizes the current state of EPA review activities, recognizing that there is no single approach. It depends on the nature of the model, its application, the needs of the model developers and users, the peer review guidance being followed, and the requirement of the specific regulatory environment statutes. For the purposes of this section, external reviews are categorized as peer review, public review, and interagency review.

Peer Review

This category refers to technical experts reviewing the model and its application for scientific merit. Although it is expected that key elements of models will be published in the peer reviewed literature, this discussion does not address journal reviews. Peer review is embedded in the history of science because of its value in improving the quality of a technical product and providing assurance to nonexperts that the product is of adequate quality. These values are so important that attention must be paid to the quality of the peer review itself and to whether the comments were addressed and appropriately incorporated into the final product. All peer reviews are not equivalent. A peer review on model code, for example, will be useful, but inadequate to evaluate the utility of the model for a specific application. Thus, the charge to each peer review for a model and its application needs to be considered relative to the criteria for model evaluation and where the model is in its life cycle, as described in Chapter 4.

In July 1994, EPA published *Guidance for Conducting External Peer Review of Environmental Regulatory Modeling* (EPA 1994c), which was a prelude to broader peer review guidance published in 2006 (EPA 2006a). The 2006 guidance is very comprehensive and detailed, describing such elements as matching the kind and degree of peer review to the impact of the work product (product of influential scientific information, very influential scientific information, or other) or rule (Tier 1, 2, or 3 rule), determining resources needed for peer review, selecting peer reviewers, documenting the review, and so forth. EPA has also created an "action development process" for regulations and other decisions. OMB also has published the *Final Information Quality Bulletin for Peer Review* (70 Fed. Reg. 2664 [2005]). Although these three documents differ in details, they are conceptually similar because they *require* peer review of models or their applications that are most likely to have "major" or "substantial" impacts. They also describe the need for peer reviewers to have the necessary technical expertise and to be free of conflicts of interest and the need for a panel to balance biases. OMB's guidance has greater emphasis on the need to make key elements of the review available to the public. The EPA Science Inventory keeps a list of the different science activities and their required levels of peer review. Its activities are broad and described at http://cfpub.epa.gov/si/si_pr_agenda.cfm.

The guidance on regulatory models calls for reviews with the goals of "judging the scientific credibility of the model including applicability, uncertainty, and utility (including the potential for misuse) of results, and not for directly advising the agency on specific regulatory decisions stemming in part from consideration of the model output" (EPA 1994c). Box 2-4 lists elements of peer review described by EPA for use with regulatory models. This guidance also offers a framework for reviewing model development, model application, and environmental regulatory decision making. It explains that policy decisions resulting from the science and other factors are required by law to be made by EPA decision makers. The policy decisions are often subject to public comment.

BOX 2-4 Elements of External Peer Review for Environmental Regulatory Models

Model Purpose/Objectives

- What is the regulatory context in which the model will be used and what broad scientific question is the model intended to answer?
- What is the model's application niche?
- What are the model's strengths and weaknesses?

Major Defining and Limiting Considerations

- Which processes are characterized by the model?
- What are the important temporal and spatial scales?
- What is the level of aggregation?

Theoretical Basis for the Model—formulating the basis for problem solution

- What algorithms are used within the model and how were they derived?
- What is the method of solution?
- What are the shortcomings of the modeling approach?

Parameter Estimation

- What methods and data were used for parameter estimation?
- What methods were used to estimate parameters for which there were no data?
- What are the boundary conditions and are they appropriate?

Data Quality/Quantity

Questions related to model design include:

- What data were utilized in the design of the model?

- How can the adequacy of the data be defined taking into account the regulatory objectives of the model?

Questions related to model application include:

- To what extent are these data available and what are the key data gaps?
- Do additional data need to be collected and for what purpose?

Key Assumptions

- What are the key assumptions?
- What is the basis for each key assumption and what is the range of possible alternatives?
- How sensitive is the model toward modifying key assumptions?

Model Performance Measures

- What criteria have been used to assess model performance?
- Did the data bases used in the performance evaluation provide an adequate test of the model?
- How does the model perform relative to other models in this application niche?

Model Documentation and Users Guide

- Does the documentation cover model applicability and limitations, data input, and interpretation of results?

Retrospective

- Does the model satisfy its intended scientific and regulatory objectives?
- How robust are the model predictions?
- How well does the model output quantify the overall uncertainty?

Source: EPA 1994c.

EPA has several forums to conduct peer reviews: the EPA Science Advisory Board (SAB), the EPA Clean Air Science Advisory Committee (CASAC), the EPA Science Advisory Panel (SAP), or ad hoc committees. They are described in more detail in Box 2-5. The first three organizations are convened under the Federal Advisory Committee Act and are subject to requirements of that act, including that all meetings and deliberations must be public. Major ad hoc committees also hold open meetings. Typically, the charges to SAB, CASAC, and SAP are broad. Ad hoc committees are often used for more in-depth reviews. All types of peer review are of substantial value, but the adequacy of peer review of a model must be judged in context with the need for evaluation of each major step from model conception to application. Major reviews, such as those performed by SAB, besides providing valuable input to agency scientists and managers, can become a part of the administrative record and can be used in court challenges. Examples of model peer reviews are the SAB reviews of the 3MRA model (EPA 2004e), the SAB review of the EPA Region 5 critical ecosystem assessment model (EPA

BOX 2-5 The Different Types of Science Advisory Panels at EPA

The CASAC was established under the CAA to review EPA's NAAQS and report to the EPA administrator. It is administratively housed in SAB. This group reviews the "criteria documents" of the criteria air pollutants to evaluate whether the information contained is adequate to support a decision. They also review the staff paper that has the EPA staff's recommendations for the standard. Both documents rely on models.

SAB traces its history to 1978. Its charge is to provide independent science and technical advice, consultation, and recommendations to the EPA administrator on the technical bases for agency positions and regulations. Most of its activities involve reviewing technical documents, including numerous model reviews (e.g., EPA 2004e, 2005b). SAB also produced the *Resolution on the Use of Mathematical Models by EPA for Regulatory Assessment and Decision-Making* (EPA 1989).

The federal Insecticide, Fungicide, and Rodenticide Act established SAP in 1975. The Food Quality Protection Act mandated a science review board of scientists who would be available to SAP on an ad hoc basis. SAP provides scientific advice, information, and recommendations to the EPA administrator on pesticides and pesticide-related issues as to the impact of regulatory actions on health and the environment. Several SAP panels have focused on models to predict exposures to pesticides or on pesticide health assessments that were partly based on models. SAP panels summarize their discussions and issue recommendations in the minutes of the meetings (e.g., EPA 2005c).

Ad hoc committees are often used by EPA when the document being reviewed does not have the impact that invokes the need for SAB, CASAC, or SAP. As related to models, they might involve highly technical reviews before the SAB-level stage or might be for risk assessments that include some degree of reliance upon models.

2005b), and the SAP preliminary evaluation of physiologically based pharmacokinetic and pharmacodynamic modeling for the *N*-methyl carbamate pesticides (EPA 2005c).

Public Review

Public review of a regulatory model concerns review and comments by stakeholders during the public comment periods of external peer review activities or during the "notice and comment" period that accompanies rule-making activities. Herein, "stakeholder" is defined as a person or nonfederal entity and external to the agency not involved in the above-described peer review. They include members of the general public. Thus, many individuals and entities are stakeholders and have different interests, capabilities, and capacities to perform this role. For example, consider the different capabilities to generate comments on models and model results between a member of the general public with limited abilities to perform computational analysis and a corporation or other organization with a substantial scientific staff. These differences need to be understood and accommodated when fulfilling the intent and actual requirements for public review. When EPA requests a peer review by CASAC, SAB, or SAP, the document is made public, and the public is able to comment at the public meetings of these organizations as per the Federal Advisory Committee Act. Furthermore, EPA is required by statute to solicit comments from affected parties and the public at large on all final proposals for agency action (5 U.S.C. § 553). A mandatory "notice and comment" process is intended to ensure that the agency informs the public of its activities and takes their concerns and input into account. According to statute, EPA must also make all relevant documents in the record supporting its decision available to the public for viewing during the comment process.

Interagency Review

EPA's regulations are developed and implemented as part of a larger federal fabric. For example, some of EPA's regulations affect other agencies directly (for example, Department of Defense Superfund sites) and indirectly (for example, economic consequences to policies of other agencies). A example of an EPA model that plays a critical role in another agency's activities is the motor vehicle emissions factor model, (MOBILE), which plays an important role in the Department of Transportation (DOT) transportation planning activities (Ho 2004). This has inspired DOT to evaluate aspects of MOBILE directly (Tang et al. 2003a,b). Thus, there is a variety of both formal and informal processes for interagency review of regulatory models and analysis based on these models. The majority of interagency reviews involve mandatory oversight by OMB, although other agencies may also engage in more informal review and comment. Under various executive directives, OMB review is generally cursory unless the regulatory program, which the model informs, is deemed to be "significant" with respect to its economic implications (Graham 2004). OMB oversees these process requirements and will work with the agencies to ensure that their regulatory analyses are satisfactory. OMB review of other agencies' rule-makings is generally established through executive order and, while these presidential directives are mandatory, agency violations cannot be enforced through the courts.

Completing the Review Cycle

Several of the processes of external reviews are still not transparent in regard to the disposition of the comments. In some instances, the effect of comments on the regulatory process is not clear. It is understood that not all comments are appropriate or useful, even though all need to be carefully considered. Thus, the issue is transparency—those commenting, from prominent scientists on the SAB to members of the general public, need to understand how their comments were considered. EPA's *Peer Review Handbook* (EPA 2006a) discusses this issue and calls for a written record of response to comments. EPA has an exemplary process in terms of transparency for the NAAQS where a public docket contains both the original comment and the agency's responses.

Legal Challenges to EPA's Models

Laws and executive orders not only provide a mechanism for increased external inputs to EPA's models but also provide opportunities for adversarial challenge. There are two formal opportunities for interested parties to challenge EPA's models. The first and most established is the ability of interested parties to challenge agency action in court. If the model supports a regulation and has been subject to notice and comment, the courts give EPA considerable deference. Thus, challenges to EPA models are successful only when the regulation (and/or underlying model) is in conflict with EPA's statutory mandate, has been determined to be inconsistent with Administrative Procedure Act requirements, or is "arbitrary and capricious" (5 U.S.C. § 706). As one court summarized in reviewing a model: "This Court must not undertake an independent review of EPA's scientific judgments; our inquiry focuses only on whether the agency has met the statutory requirement for 'sufficient evidence.'" (National Oilseed Processors Ass'n v. Browner, 924 F. Supp. 1193, 1209 [D.D.C. 1996], affirmed in part and reversed in part on other grounds, Troy v. Browner, 120 F.3d 227 [D.C. Cir. 1997]). If the model has not been subject to notice and comment but creates obligations for private parties—for example, at the permitting or enforcement stage—those affected by the model can typically challenge either the model or its application in court. In some of the cases, the agency may receive much less deference from the courts compared to the situation where the model has been subject to notice and comment (for example, see United States v. Plaza Health Laboratories, Inc., 3 F.3d 643 [2d Cir. 1993]) (applying the rule of lenity, rather than deferring to EPA, in interpreting "point source" in a criminal CWA prosecution). Generally, a complete model history documenting the justification for various decisions related to model design and development may help the agency defend a model against formal challenges.

EPA's models have sometimes been challenged, and in some cases, challengers have been successful in forcing the model or its application back to EPA to correct what the courts view as fundamental flaws. Some of this judicial activity may be a result of EPA's past, ad hoc approach to developing and using models; a more rigorous and formalized approach might ward off some of these challenges by instituting more rigorous modeling practices in the agency. For example, when EPA declines to explain its decision or revise a supporting model even after receiving comments refuting one of the model's critical assumptions, the courts have invalidated and remanded the model back to EPA. Challengers have also been successful when they establish that EPA's model is not applicable to a particular subset of industries, activities, or locations. If EPA applies a generic air dispersion model to a large power plant located in a meteorologically unusual setting, such as the shores of Lake Erie, EPA might have to test the location to establish that the model provides some reliability in that setting, or it must be prepared to explain why its model should be accepted as is (for example, State of Ohio v. EPA, 784 F.2d 224 [6th Cir. 1986] and 798 F.2d 880 [6th Cir. 1986]).[8] Finally, if challengers disagree with embedded policy judgments, such as the risk adversity of assumptions built into a risk assessment, courts will sometimes invalidate a model and not defer to the agency (Gulf South Insulation v. Consumer Product Safety Commission, 701 F.2d 1137 [5th Cir. 1983]). However, this line of cases is more complex and unpredictable (Pierce

---

[8]Remanding EPA's air dispersion model because EPA had not adequately demonstrated that its CRSTER model took into account the "specific meteorological and geographic problems" of the modeled large power plants situated on the shores of Lake Erie. It was therefore arbitrary and capricious for EPA to allow a 400% increase in emissions "without evaluation, validation, or empirical testing of the model at the site."

1988).[9] Because these legal challenges are time-consuming and costly, they are typically mounted only when an affected or interested party stands to gain something important—whether it is gaining less stringent regulatory requirements or positive publicity for members—from a challenge.

A second, more recent opportunity for external challenge to model use in the regulatory process is through the Information Quality Act (Treasury and General Government Appropriations Act for Fiscal Year 2001, Pub. L. No. 106-554, § 515, 114 Stat. 2763 [2000]), which is implemented through OMB's *Guidelines for Ensuring and Maximizing the Quality, Objectivity, Utility, and Integrity of Information Disseminated by Federal Agencies* (OMB 2001). Some of challenges under the Information Quality Act result from EPA's occasional ad hoc approach to developing and using models. This statutory provision allows any interested person to file "requests for correction" on "information" that is "unreliable" or lacks other qualities, such as objectivity or integrity. To date, courts have refused to review these challenges, but the challenges can be appealed inside the agency and the agency must respond to complaints that the information, including information used in models or the models themselves, is unreliable. However, there are continued efforts to make challenges under the Information Quality Act reviewable by the judiciary (Shapiro et al. 2006).

Challenges filed under the Information Quality Act to date generally target technical decisions within EPA that have important economic consequences (EPA 2006i). In at least one instance—the Competitive Enterprise Institute's (CEI's) challenge to the climate change models used in the National Assessment on Climate Change—the challenge has been directed specifically at agency models (EPA 2003c). In the case, CEI argued that the models were not reliable and had not been adequately peer reviewed. The agencies denied the petitions and CEI's internal appeals. CEI then appealed its case to the D.C. District Court where CEI ultimately withdrew its case. Information Quality Act challenges brought by affected parties sometimes seek correction of flaws or technical misstatements in agency documents, but in other instances, as in the Competitive Enterprise Institute's challenge, they have requested that the agencies cease dissemination of the information. If an agency

[9]Arguing that judges on the D.C. Circuit may be substituting their own interpretations of ambiguous statutes for agencies and randomly reversing agency policy making in rule-makings.

denies a petition on appeal, as has been the case for most IQA challenges filed to date, the challenge fails.

THE CHALLENGES OF MODELING IN A REGULATORY ENVIRONMENT

This chapter has described the types of models used in EPA's regulatory activities, how models fit into the regulatory process, and legal and other constraints governing their use. Modeling is a difficult enterprise even when it is not being conducted in an adversarial regulatory environment. Further, the range of model applications is vast, and many agencies and stakeholders are involved in producing analysis. When the demands of regulatory accountability, transparency, public accessibility, and technical rigor are added to the challenges typically encountered in modeling, the task becomes much more complex.

Although improvements to EPA regulatory modeling efforts are possible, EPA clearly has made important advances in the science of environmental modeling and has been a global leader in using models in the environmental regulatory decision process. However, future regulatory modeling activities will be challenged by new scientific understandings, expanding sources of environmental and human observations, and new issues. To meet the challenges, continued improvement in model practices will be required. In this chapter, the committee offers recommendations related to continuing improvements to the accessibility of regulatory modeling. Later in this report, we offer recommendations related to model evaluation; principles for model development, selection, and application; and model management.

Model Goals

Models are used in regulatory settings when EPA determines that a model will be useful in reaching or enforcing a regulatory decision. Given the diversity of regulatory aims and targets, however, a wide variety of models and modeling goals can exist. At one extreme, the agency can use a model that provides the best technical analysis of the concentrations of ambient air pollutants and resulting health and environmental impacts most likely to result from combined industrial and nonindustrial emissions controls. This precision is desired because of the enormous compliance costs associated with emissions controls and the enormous health costs if air pollutants are not correctly estimated and exceed allowable levels. At the other extreme, EPA might want to use a model that provides only a crude and inexpensive prediction for a system.

The regulatory environment also creates the opportunity for many different types of legal constraints on modeling that are foreign in non-regulatory settings. Congress may instruct, for example, that a regulation err on the side of over-predicting public health harms. Other constraints might result from legislative mandates that EPA develop and use models in situations where resources, including both time and financial support, are scarce.

Time and resource limitations can also lead EPA to use existing models outside their "application niche," a set of conditions for which the model is designed to be useful. There is some evidence, for example, that EPA and other agencies have sometimes used a model in a setting where the model no longer provides useful guidance. For example, EPA's generic test to predict the toxicity of wastes in landfill settings (the Toxicity Characteristic Leaching Potential Test) generally adopts worst-case assumptions. Yet, in some disposal settings, the worst-case assumptions have been challenged successfully as inapplicable for specific types of disposal operations, such as for the disposal of a particular type of waste (potliner waste) in a monofill (for example, see Columbia Falls Aluminum Co. v. U.S. Environmental Protection Agency, 139 F.3d 914 [D.C. Cir. 1998]; Edison Electric Institute v. U.S. Environmental Protection Agency, 2 F.3d 438 [D.C. Cir. 1993]; and Association of Battery Recyclers, Inc. v. U.S. Environmental Protection Agency, 208 F.3d 1047 [D.C. Cir. 2000]).

Technical Reliability

The sometimes contentious environment for regulatory models also creates important impediments for ensuring the technical reliability of EPA's models. Formal evaluation processes required by administrative law may deter meaningful model reevaluation and adjustment over time. Once a regulatory action has survived the multilayered review and challenge processes, it may remain in place for some time. Indeed, rule-making requirements can be read to require that the agency undergo notice and comment and the risk of judicial review every time it revises a model that supports a rule-making, since it must ensure that there has been "meaningful public comment" on all aspects of its final rule (for example, see Small Refiner Lead Phasedown Task Force v. EPA, 705 F.2d 506, 540-41[D.C. Cir. 1983]). This inertia is not ideal for any regulatory decision, but it is particularly unfortunate for models. The cumbersome regulatory procedures and the finality of the rules that survive them are directly at odds with the dynamic nature of modeling and the goal of improving models in response to experience. Although some stakeholders may prefer a constant model because of the stability it provides, this model might not reflect the most updated science.

Transparency and Accountability

In the regulatory environment, EPA has the responsibility to ensure that a model's development and use is transparent. Because modeling is often a very technical exercise, EPA faces a challenge in making all of the underlying decisions intertwined within a model intellectually accessible to a nontechnical audience. A model that attempts to determine the fate of a chemical in soil, for example, may involve choices between competing assumptions, such as the percolation rate of a chemical at a particular location. Selection of the most appropriate assumption in some cases may depend not only on technical judgment but also on the policy goals of the modeling effort. A recent EPA report documents how science mingles with policy in health risk assessment (EPA 2004a). If the model is supposed to err on the side of protecting health and the environment, the model may need to err on the side of quicker percolation rates when several choices are plausible. Making these choices explicit and accessible is a challenge because policy judgments can be numerous and varied in their importance. Nevertheless, administrative processes expect EPA to make many of these types of judgments and technical decisions transparent so that affected stakeholders and the general public can comment on the model and its regulatory implications.

Because models are uncertain and are used to make policy, stakeholders necessarily play a vital role in EPA's development, use, and evaluation of models. Differing interpretations of risk, risk preferences, and a range of other values and understandings mean that a broad array of participants will have much to add to the modeling exercise. As a result, these various constituencies and individuals must be able to participate in the model evaluation process through various activities, including producing their own supporting or conflicting model results, and challenging the legitimacy or accuracy of a model in public comments or judicial actions.

Clearly, EPA faces many difficult challenges in making its models, particularly its complex models, accessible to the diverse interests. Nevertheless, EPA has taken a major step in the right direction through the CREM database of models. This information further enhances the transparency and understandability of models to a wide array of interested participants. Despite these efforts, however, stakeholders with limited resources or technical expertise still face substantial barriers to being able to evaluate EPA's models, comment on important model assumptions, or use the models in their own work.

Recommendations

EPA should place a high priority on ensuring that stakeholders and others have access to models for regulatory decision making. To ensure that its models database contains all actively used models, EPA should continue its support for the intra-agency efforts of CREM. A more formal process may be needed to ensure that CREM's models database is complete and updated with information that is at least equivalent to information provided for models currently contained in the database.

Yet, even with a high-quality models database, EPA should continue to develop initiatives to ensure that its regulatory models are as accessible as possible to the broader public and stakeholder community. The level of effort should be commensurate with the impact of the model use. It is most important to highlight the critical model assumptions, particularly the conceptual basis for a model and the sources of significant uncertainty. Meaningful stakeholder involvement should be solicited at the model development and model application stages of regulatory activity, when appropriate. EPA could improve model accessibility through a variety of activities, such as requiring an additional interface for each model to help to identify the assumptions and sources of parameters and other uncertainties and providing additional user and stakeholder training.

However, even if full information on a model is available, technical expertise will still be required to judge independently its quality and suitability for regulatory application. Each of these recommendations requires staff time and resources, which may be considerable. Thus, EPA's efforts to enhance opportunities for public participation in any particular case must be balanced against other agency priorities.

3

Model Development

INTRODUCTION

Models are a simplification of reality that can be compared to maps. Road maps indicate certain aspects of reality (for example, roads of a certain size) and not others (for example, sewer lines, power lines, and buildings). No one map can include all aspects of reality and, similarly, all models, no matter how complex, are constrained by basic assumptions, structure, and uncertainties. Model development involves the definition of model objectives, conceptualization of the problem, translation into a computational model, and model testing, revision, and application. Although almost all model development follows these general steps, models designed for regulatory purposes are subject to constraints in addition to those for models developed strictly for research. This chapter focuses on how model development might best proceed toward regulatory objectives, although there is no one route for successful model development. Our objective is not to provide a treatise on model development. Many other references in an array of disciplines provide comprehensive descriptions of model development for various types of models (Starfield and Bleloch 1991; Clemen 1995; Mesterton-Gibbons 1995; Beck 2002a; Bassetti and Woodward 2005; Ramaswami et al. 2005). This chapter discusses the major steps in regulatory model development focusing on the main lessons learned from previous efforts in EPA and other organizations. It is intended to discuss some of the literature on model development and provide a general framework for EPA as it goes about its business. The wide range of environmental model types makes our effort prone to both overgeneralization and oversimplification. To reduce such difficulties, we will often refer to examples of regulatory model development, especially those from air quality modeling. Box 1-1 in Chapter 1 contains a brief history of EPA's effort to model tropospheric ozone.

ALTERNATIVE MODEL DEVELOPMENT PATHWAYS

Some regulatory models arise from those developed as general research tools. Others were developed specifically for addressing regulatory issues. They have been developed by EPA scientists, academia, national laboratories, or the private sector. Some of the most complex models have benefited from contributions by almost all of the above. For example, complex regional chemical transport models for simulating air quality usually include components contributed by multiple parties. The urban airshed model (UAM), heavily used for the design of ozone control strategies in the 1980s and 1990s, was developed by a private company (Systems Applications International) relying on contributions of academia and on support from public and private organizations. The major air quality model developed for use in-house by EPA is the community multiscale air quality (CMAQ) model (EPA 1999b). EPA and NOAA scientists developed the most recent CMAQ model in partnership with a nonprofit organization (Microelectronics Center of North Carolina) and contributions by academia funded by EPA, the National Science Foundation and state authorities, most prominently California authorities (CMAS 2006). A variation of CMAQ, called CMAQ-MADRID, has been developed by a private company (Atmospheric and Environmental Research, Inc) using the CMAQ model as a starting point and adding components developed by academic researchers or by company scientists (Zhang et al. 2004). A private organization, Electric Power Research Institute, provided funding for the CMAQ-MADRID development. All the above codes are in the public domain.

Under the Toxic Substances Control Act, EPA must make individual pre-manufacturing decisions on 2,000 new chemicals per year before a new chemical can enter the market. Because of the large number of decisions, the agency has had to rely on screening tools that predict properties from chemical structure. EPA uses EPI (Estimation Programs Interface) Suite, which consists of several quantitative structure-activity relationships (QSARs) models that are available in the public domain.

However, the set of models began as a few proprietary models developed by Syracuse Research Corporation. Later some of the models were developed in collaboration with EPA, and then all the models were sold to EPA. QSARs are able to take complex chemical structures and predict physical properties, behavior in the environment, and toxicity (Jaworska et al. 2003; Tunkel et al. 2005). The EPI Suite is used by EPA's Office of Pollution Prevention and Toxics to predict physical-chemical properties, environmental fate and transport, and aquatic toxicity for regulatory decisions on new chemicals when data are not available. The models are also used by industry for pollution prevention and by many government agencies for identifying persistent, bioaccumulative, and toxic (PBT) chemicals (Jaworska et al. 2003).

Alhough the development paths of models may be different, many end up having long lives in the regulatory process. Table 3-1 shows the life history of the MOBILE model, which is used to estimate atmospheric emissions from vehicles. This table indicates the periodic revisions that necessarily accompany a model that has been in use for almost 30 years. In the case of MOBILE, such revisions are often major overhauls and updates of the model, resulting in emissions estimates being much different from these in previous versions (NRC 2000; Holmes and Russell 2001). Along with the UAM discussed above, the QUAL2 water quality model is an additional example of a regulatory model that has seen multiple versions and major scientific modifications and extensions in over 2 decades of existence (Barnwell et al. 2004).

OVERVIEW OF MODEL DEVELOPMENT

Jakeman et al. (2006) separates model development and evaluation into the 10 steps shown in Figure 3-1. The committee agrees with the concept shown in Figure 3-1 that model development is typically an iterative process, especially for long-lived models used over several decades. However, for the purposes of this chapter, the model development process is compressed into the six phases shown in Box 3-1. Documentation occurs at each step of the process, as do certain aspects of evaluation. Chapter 4 describes in detail the evaluation process that occurs throughout the model's life cycle, compressing the model lifecycle further into 4 steps (problem identification, conceptual

TABLE 3-1 MOBILE Model Revisions

Version	Release Date	Model Revisions
MOBILE1	1978	• Included modeling of exhaust emissions rates as functions of vehicle age and mileage (zero-mile levels and deterioration rates).
MOBILE2	1981	• Updated with substantial data (available for the first time) on emission-controlled vehicles (catalytic converters, model years 1975 and later) at higher ages and mileages. • Provided additional model user control of input options.
MOBILE3	1984	• Updated with substantial new in-use data. • Elimination of California vehicle emissions rates (continued to model low- and high-altitude emissions). • Added tampering (rates and associated emissions impacts) and antitampering program benefits. • In-use emissions-factor estimates for nonexhaust emissions adjusted for real-world fuel volatility as measured by Reid vapor pressure (RVP).
MOBILE4	1989	• Updated with new in-use data. • Added running losses as distinct emissions source from gasoline-powered vehicles. • Modeled fuel volatility (RVP) effects on exhaust emissions rates. • Continued expansion of user-controlled options for input data.
MOBILE4.1	1991	• Updated with new in-use data. • Added numerous features allowing user control of more parameters affecting in-use emissions levels, including more inspection and maintenance (I/M) program designs. • Included effects of various new emissions standards and related regulatory changes (for example, test procedures). • Included impact of oxygenated fuels (for example, gasohol) on CO emissions.
MOBILE5	1993	• Updated with new in-use data, including basing new basic emissions-rate equations on much larger database derived from state-implemented IM240 test programs. • Included effects of new evaporative emissions test procedure (impact on in-use nonexhaust emissions levels).

MOBILE5	1993	• Included effects of reformulated gasoline (RFG). • Included effects of new NO_x standard of 4.0 g/bhp-hr for heavy-duty engines. • Included impact of oxygenated fuels on VOC emissions. • Included Tier 1 emissions standards under 1990 Clean Air Act Amendments. • Added July 1 evaluation option. • Included impact of low-emission vehicle (LEV) programs patterned after California regulations. • Revised speed corrections used to model emissions factors over range of traffic speeds.
MOBILE5a	1993	• Corrected a number of minor errors in MOBILE5.
MOBILE5b	1996	• Included final on-board vapor-recovery regulations. • Included final reformulated gasoline regulations. • Added more user options for I/M programs.
MOBILE6	2002	• Added the effects of Tier 2 and new heavy-duty engine and diesel fuel rules. • Updated with new and improved data in many areas, including in-use deterioration of 1981 and newer vehicles, light-duty speed effects, gasoline sulfur effects, and evaporative emissions. • Revised I/M benefits algorithm; removed calculation of purge test benefit. • Revised algorithms for air conditioning and high acceleration driving. • Expanded number of vehicle subclasses from 8 to 28. • Added hourly calculation of emissions and emission estimates by roadway type. • Separated start and running exhaust emissions.
MOBILE6.2	2004	• Added ability to model emission factors for particulate matter and six air toxics; added ability to model additional air toxics with user-supplied emission factors. • Updated carbon monoxide emission factors.

Source: EPA 1999c, 2006j.

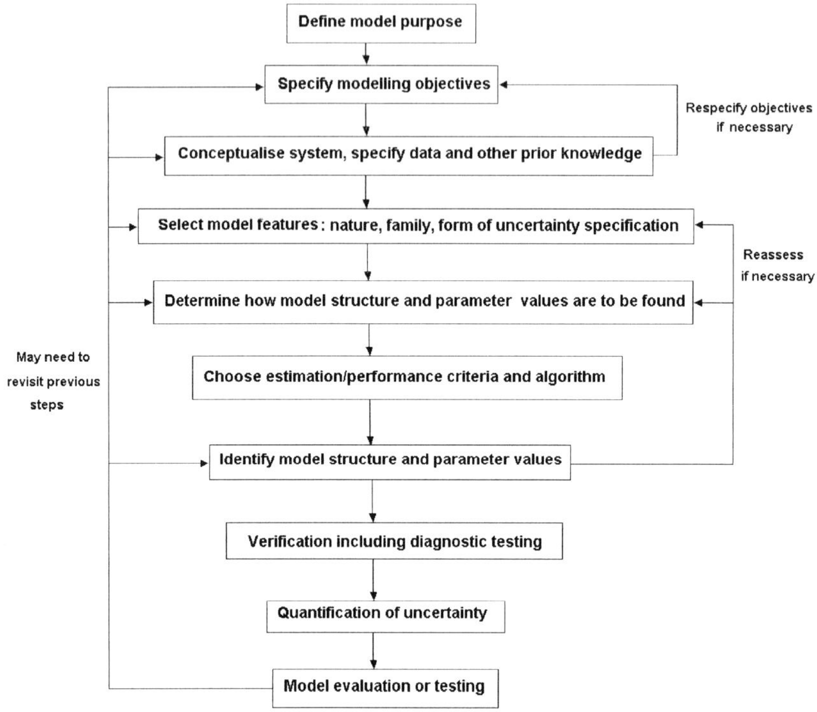

FIGURE 3-1 Iterative steps in model development proposed by Jakeman et al. (2006). Source: Jakeman et al. 2006. Reprinted with permission; copyright 2006, *Environmental Modelling & Software.*

model development, computational model development, model use) to make the evaluation process more tractable to the reader.

A general issue concerns the uses for which a model is being constructed. This report includes models that are used for two main purposes: those used before regulations are developed to strategically plan and assess priorities and design, evaluate, and propose regulatory approaches (hereafter referred to as pre-regulatory planning models) and those used to implement regulatory programs, including programs that have been delegated to states and local governments (hereafter referred to as post-regulatory implementation and compliance models). These are two quite separate uses, although some models can be used for both purposes as discussed in Chapter 2. However, there might be some important differences. Pre-regulatory planning models require a more general

BOX 3-1 Basic Steps in Modeling Development Process

Model Development Step	*Modeling Issues*
Definition of Model Purpose	Goal Decisions to be supported Predictions to be made
Specification of Modeling Context	Scale (spatial and temporal) Application domain User community Required inputs Desired output Evaluation criteria
Conceptual Model Formulation	Assumptions (dynamic, static, stochastic, deterministic) State variables represented Level of process detail necessary Scientific foundations
Computational Model Development	Algorithms Mathematical/computational methods Inputs Hardware platforms and software infrastructure User interface Calibration/parameter determination Documentation
Model Testing and Revision	Theoretical corroboration Model components verification Corroboration (independent data) Sensitivity analysis Uncertainty analysis Robustness determination Comparison to evaluation criteria set during formulation
Model Use	Analysis of Scenarios Predictions evaluation Regulations assessment Policy analysis and evaluation

framework, allowing alternative policy initiatives to be analyzed, perhaps by varying basic model assumptions. This analysis may be done at the national scale and by EPA. Post-regulatory implementation and compliance models will typically be more closely tied to site-specific observational data, producing a plan for implementing a regulation or assessment of compliance for a given location or substance. Besides being used by EPA, modeling in the post-regulatory process may also be done by state and local governments and their consultants.

INTERDEPENDENCE OF MODELS AND DATA FROM MEASUREMENTS

Developing and evaluating models typically requires dependence on measurements. In some cases, there are plenty of measurement data for developing model parameters, boundary conditions, and other inputs. Often, however, data are missing, which is an inherent factor in the need for models. Optimally, measurements and models develop iteratively, each informing the other. Box 3-2 describes some examples where measurement data have influenced model development. Although there are trade-offs about whether it is preferable to invest in more data or in better models, the committee does not conclude that the problem of resource allocation for data versus models can be viewed simply as an optimization problem. The difficulty in attempting to formulate such a problem is how to define the optimization criteria and objective function —how does one define and represent the benefit from additional data one does not have relative to investing in additional modeling one does not have. Further, data are typically collected to fulfill multiple objectives, including determining compliance with environmental regulations, further complicating attempts to formulate the data versus model issue into an optimization context.

MODEL DEVELOPMENT PHASES

Definition of Model Purpose

The first step involves defining the major purpose or purposes for which the model is developed. As discussed in Chapter 4, this occurs at the problem identification stage when decision makers, model developers, and other analysts must consider regulatory needs and whether

BOX 3-2 Interdependence of Models and Data from Measurements

Models are developed and evaluated using a wide range of data, theories, and assumptions and are revised in the process. The wisdom of iteration of measurements and modeling is illustrated by three examples.

Persistent organic pollutants (POPs) in the Arctic. POPs are chemical substances that persist in the environment, bioaccumulate through the food web, and pose a risk of causing adverse effects to human health and the environment. The first evidence for long-range transport of these substances came about when measurements in animals and the environment of the Arctic revealed the presence of POPs that were never produced there. The lack of reliable emissions data led to a number of modeling efforts used to explore hypotheses regarding the atmospheric transport of and deposition of POPs in the Arctic. For example, Wania and Mackay (1995, 1999) introduced multimedia global distribution models for persistent organic chemicals with a focus on transport and deposition to the Arctic. Then Scheringer (1996, 1997) developed evaluative models to assess global persistence and spatial range as end points in screening level assessments. These models and their results provided key insight both to international agencies, such as the United Nations, and to innovative scientists working independently to measure how POP concentrations vary with latitude. These new measurements provided important feedback that made it possible to develop the next generation of models by merging results from both the first generation of models and the new measurements.

Pharmacokinetic modeling. Andersen et al. (2005) describe examples of how integrated measurements and modeling have advanced risk assessment modeling by providing more insight on how intake of chemicals by humans relates to tissue dose and metabolism. Early pharmacokinetic models of the time course of absorption, distribution, metabolism, and excretion of chemicals relied on concepts buttressed by rudimentary information. By the 1950s, data on tissue volume, blood flow, and metabolic pathways were emerging, resulting in early physiologically based pharmacokinetic (PBPK) models. These models, in turn, led to the identification of key input variables (for example, blood flow through various tissues and metabolic parameters), the measurement of which would advance the models. The first use of a PBPK model in a formal risk assessment was for dichloromethane in 1987. Advances were made in assessment methods (for example, EPA's reference concentration method) as well as in PBPK models of specific chemicals (acrylic acid, vinyl chloride, and dioxin). This iterative process continues today to inform risk assessments that can be used in regulation. It also provides a platform for more novel computational and biological systems approaches of the future (Anderson et al. 2005).

Comprehensive Everglades Restoration Plan. The planned restoration of the Florida Everglades is the largest ecosystem restoration effort ever undertaken in terms of it geographical extent and number of individual components. The NRC Committee on Restoration of the Greater Everglades Ecosystem, which was charged with providing scientific advice on this effort, describes the role that modeling and measurements should play in implementing an adaptive approach to restoration (NRC 2003). Under the committee's vision, monitoring of

(Continued on next page)

hydrological and ecological performance measures should be integrated with mechanistic modeling and experimentation to better understand how the Everglades functions and how the system will respond to management practices and external stresses. Because the individual components of the restoration plan will be staggered in time, the early components can be used as experiments to provide scientific feedback to guide and refine implementation of later components of the plan.

modeling could contribute to the regulatory process. If there is sufficient need for computational modeling, modelers must work with decision makers to define the goal of the model, the decisions it supports, and the groups that might use the model.

Addressing these questions is important for setting the direction of the model. As described in Chapter 2, legislative, regulatory, or policy mandates will often drive model development and implementation. For example, a legislative or policy mandate may require that EPA protect the most exposed individual, the most vulnerable individual, or reasonably highly exposed individual and that the agency consider long-term average exposure, the highest one-day exposure, the most exposed subpopulation, or the location of highest concentration. Indeed, legislative mandates may sometimes force development of new models or require major modifications to existing ones. This initial stage sets the direction for the conceptual model and the computational model development. The sidebar from *Alice in Wonderland* illustrates this message. If you do not know where you want to go, it may appear to others that the direction you take is not particularly important. The key goal of this initial phase is to identify whether modeling would be an effective tool for the problem at hand.

Potential uses of an environmental model include the following (Jakeman et al. 2006):

- Long-term prediction (both extrapolating from the past and answering "what if" questions).
- Short-term forecasting.
- Interpolation (estimating variables that have not or cannot be measured directly).
- Concise summarizing of data.
- Data assessment (coverage, limitations, inconsistencies, and gaps).

- Control system design (monitoring, diagnosis, decision making, and action taking).

Regulatory models are also used to do the following:

- Help determine compliance with a particular regulation.
- Evaluate a variety of alternative regulations.
- Provide a general framework to assess compliance with multiple regulations.
- Summarize available knowledge needed for regulatory decisions.

Insight from Alice's Adventures in Wonderland by Lewis Carroll

Alice speaking to the Cheshire cat: "Would you tell me, please, which way I ought to go from here?"
Cheshire cat: "That depends a good deal on where you want to get to."
Alice: I don't much care where."
Cheshire cat: "Then it doesn't matter which way you go."

Even if defining the model purpose appears to be a straightforward and easy step, it is often difficult to be clear about the purposes of an environmental model and its application domain. For scientists, the major objective is often to describe the processes dominating the behavior of the system, and for a decision maker, the objective might be to provide clear assessments of policy options. These motives are not mutually exclusive, but neither do they overlap completely. In addition, policy makers may want results for policy variables not directly represented in the model, or they may want results at scales not easily represented by a model. In any case, it is important to establish clearly the purpose and priorities of the specific model.

Specification of Modeling Context

After determining the purposes of the model, the modeler must develop specifications for the model context. This task involves addressing such questions as

- At what temporal and spatial scales is the model to be applied? This question involves the grain (resolution in time and space) and the extent (spatial and temporal domain) at which the model is to be focused.
- Who will be the major model users and what constraints does that imply for model application once developed? What is the level of expertise of the proposed users?
- What type of input data must the model users provide? How can these data be obtained (from other models and measurements)?
- What sources of data are available to support model evaluation?
- What are the basic outputs needed and must they be constrained by a deterministic approach or is a probabilistic approach allowable? What additional outputs, although not strictly required, might be useful to enhance model transparency (for example, ability to explain it to various stakeholders and users) and flexibility (for example, capacity for the model to be modified and applied to situations for which it was not constructed)?
- What level of reliability is required?
- What evaluation criteria should be applied to determine the applicability of the model or of particular model components?

For example, when developing a cancer health assessment of a chemical, considerations include whether to use a linear or a nonlinear model. If the latter is chosen, the model specifications will need to be based on interpretation of the mode of action. A second example is the assessment of human exposure to mobile-source emissions of particulate matter. Here the model developers must work with others to determine whether the objective is to estimate cumulative exposure, time history of exposure, peak exposure, or another measure of exposure. Model developers and others must consider whether they need to consider particle mass, particle number, and particle volume as a metric of exposure. They must also consider the spatial and temporal resolution in the data and parameters that probably will be available for the model. Finally, if the goal is to create linkages to broader health assessments of particulate matter, they must make decisions on whether to consider mobile-source contributions only or mobile sources as a component of all sources of airborne particulate matter.

The mismatch between data needed by the model and data available to the model often results in failure of the model exercise, even if the model itself may be an accurate representation of the science governing the behavior of the specific system. For example, air quality models require two major types of input: weather fields and emissions inventories. This input can be large (gigabytes of information) and impossible to obtain exclusively from measurements, so meteorological and emissions models estimate the input data and prepare the corresponding input files. Ideally, the input to the meteorological and emissions models would be based on actual measurements (for example, wind speed and direction in specific locations, vertical profiles of atmospheric properties, vehicle activity patterns, and emissions factors). Often, however, these models must use default inputs (for example, those based on emission factors from other parts of the country or even the world without taking into account the local conditions). Improving model inputs with measurements can be costly, and especially for emissions, measurement costs may overwhelm the actual modeling cost. Dealing with the issue of how to obtain the required inputs before developing the computational model and before building bridges with the measurement communities can make a substantial difference in the success of the modeling effort.

All the above questions apply to both pre- and post-regulatory models. Some specific questions probably will arise for each model, including the following:

For pre-regulatory planning models:

- What range of plans and scenarios must be considered?
- What array of impacts is to be included in the assessments of alternatives?

For post-regulatory planning and compliance models:

- Is the required decision a "bright line" compliant/noncompliant one or is a broader view (for example risk of noncompliance) allowable?
- What constraints are there on computational complexity? Will users insist on rapid assessments from the model (for example, does this need to be available in field situations) that preclude more advanced computational equipment?

Conceptual Model Formulation

A conceptual model formulates the basic model organization,

sometimes expressed graphically without the details of individual model components or assumptions. As discussed in Chapter 1, such an abstract representation provides the general structure of a system and the relationships within the system that are known or hypothesized to be important. Figure 3-2 provides an example of a conceptual model for assessing eutrophication. This conceptual model can be viewed as a map summarizing the structure of a model, the inputs, the state variables and outputs, and possibly the domain of applicability. Indeed, one of the critical roles of the conceptual model is to provide a visual description to decision makers, stakeholders, and interested parties of the model, including the fundamental relationships within the model and how inputs lead to outputs. Determination of an appropriate conceptual model relies first upon the problem formulation decisions discussed above as well as the decisions on the following:

- What basic scientific principles are involved in the model (for example, areas of physics, chemistry, and biology that need to be considered on the basis of the objectives)? Is there agreement about these principles or does their inclusion potentially result in controversy (in which case, allowing for alternative assumptions might be necessary)?
- Is an appropriate model formulation already extant?
- What level of aggregation is appropriate to the model objectives? This question applies to the scales for the model (for example, spatial and temporal averaging may be needed) and the structure of model components (for example, including demographic structure or putting all ages into a single class).
- What are the variables for the model (for example, what will it explicitly track in characterizing the system) and how are they related (often characterized by a box and arrow diagram, flow chart, influence diagram, or similar graphic)?
- What are the means by which the variables will be expressed? Are they deterministic (discrete, continuous, nominal), stochastic (discrete, continuous, nominal), and spatially or temporally dependent or static?
- What level of mechanistic detail is needed (for example, processes operating at what level should be included: cell, tissue, individual organism, population, and so forth)? Is a purely empirical approach (for example, a data-driven model, including many statistical ones) appropriate? Is a mixture of these necessary?

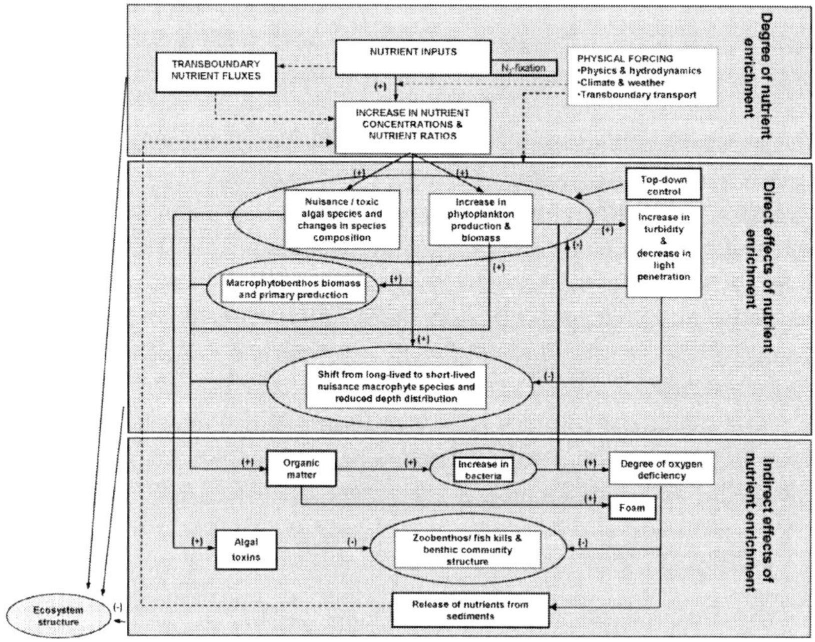

FIGURE 3-2 Conceptual model for assessing eutrophication in the European seas linking nutrient enrichments and its direct and indirect effects in the ecosystem. Source: EC 2004. Reprinted with permission; copyright 2004, the European Commission's Joint Research Centre.

- What are the model inputs and the scales at which the inputs will be provided?

There are distinct trade-offs in model development that should be addressed at the time of the conceptual model formulation. No one model can do everything. Development of a more comprehensive model will not necessarily resolve or even reduce all uncertainties in understanding and in predicting how a system will react. It is at this stage of model development that constraints, assumptions, and acceptability criteria should be established. Given financial or effort limitations, it is appropriate to set "stopping" criteria for when to decide that a model is sufficiently useful to be applied, even while acknowledging its limitations.

All the above trade-offs apply to both pre- and post-regulatory models. Some specific questions probably will arise for each model, including the following:

For pre-regulatory planning models:

- How are alternative plans formulated or specified? Do they arise from modifications of a single plan (say, by varying constraints on structures allowable to be built) or are they chosen from a broad array of options?
- What metrics are applied to compare and contrast alternative plans?

For post-regulatory implementation and compliance models:

- What criteria determine compliance versus noncompliance and how do they relate to the model state variables?
- What level of detail is needed in applications involving regulatory implementation?

Computational Model Development

This stage requires formulating the model explicitly by translating the model assumptions from the previous step into a mathematical formulation, by determining the detailed structure of the model, and by encoding the resulting model. This stage requires decisions about the following:

- Model equations that determine the relationships between variables (rules, statements, equations, statistics) and account for the mathematical structure of the model (for example, static, dynamic, discrete, continuous).
- Parameter estimations (from either data or underlying scientific assumptions) to determine input model parameters or distribution of such parameters in the case of a stochastic formulation.
- Appropriate software design and engineering tools to encode and/or solve the model, appropriate computational algorithms, and appropriate model interface to ensure applicability by the user community.
- Methods for analysis of model results, including graphic outputs and the capability to conduct sensitivity and uncertainty analysis.

- Flexibility to modify model structure and inputs in the future as new data arise, alternative objectives are specified, or different regulations are assessed.
- Documentation to allow for transparency of the model based on the needs of the user community and the potential for future modification. Such documentation maintains the history of major revisions of the model.

One critical issue is whether to revise an existing model or to develop a new one. "Model recycling" can save a huge development effort by applying a tested model to a purpose that is different from its original one. Furthermore, modelers often face the difficult decision between the development of one model that describes everything (the "swiss army knife" of models) and can be used for a variety of purposes and the development of multiple smaller models that have a common core but are developed separately for different purposes.

For example, the main purpose of the CMAQ model is to simulate concentrations of fine particulate matter and ozone in the lower atmosphere and to assist the analysis of the corresponding regulations. As a complex model, it describes the concentrations of more than 100 air pollutants in space and time. It has become a family of models (for example, CMAQ-MADRID and CMAQ-Hg) addressing a range of different air quality problems, including visibility reduction and acid deposition. Given that the different versions of CMAQ take advantage of the core of the model (atmospheric transport, gas-phase chemistry, and so forth) without violating any of the major assumptions of CMAQ, the strategy is a good one. For example, CMAQ has been extended (after some nontrivial modifications in its code) to address mercury (CMAQ-Hg). Although there are many gaps in our scientific understanding of the corresponding problem, CMAQ is an appropriate platform for such an extension. On the other hand, the model would require major redevelopment to address potential regulations of ultrafine particles (diameter less than 100 nanometers [nm]) due to numerical issues with its description of the particle-size distribution. Thus, there are limitations to the degree that CMAQ can be adapted.

Another example is the MOBILE model (Table 3-1), which has evolved from a tool for estimating regional motor-vehicle emissions inventories to a model used for estimating emissions on individual highway segments where instantaneous operating conditions of individual vehicles may be critical but that are not represented in the model. As concluded by the NRC (2000), the farther MOBILE's applications deviate from its original purpose of estimating aggregate regional emissions, the more difficult it becomes to verify the accuracy of its predictions. Because of the difficulty in developing a single motor-vehicle emissions model appropriate for all applications, the NRC (2000) recommended that EPA develop a toolkit of models based on a consistent data set and model interface. Such a toolkit would include an aggregated regional emissions component, a smaller scale model for simulating emissions along major highway corridors, and a microscale instantaneous emissions-modeling component for more transient and localized traffic conditions. For a toolkit approach, the type of motor-vehicle emissions model applied could better meet the characteristics of the problem while being consistent from one problem scale to another. Tierney (2004) described how EPA's new mobile-source emissions model, known as the MOVES model, will address this and other issues raised by the NRC.

Additional considerations probably will arise for pre- and post-regulatory models.

For pre-regulatory planning models:

- Provide methods to compare and contrast the implications of alternative plans, perhaps requiring the capability for exploratory analysis by the users of model outputs.
- Provide methods either to vary the constraints on plans-scenarios or to vary the metrics to evaluate each plan.
- Provide automated optimization methods to specify the highest ranked plan from a given set based on chosen criteria.

For post-regulatory implementation and compliance models:

- Provide methods to modify inputs to determine how readily a noncompliant case might become a compliant one and vice versa.
- Provide methods to ascertain the impacts of additional data on model results and assist users in determining the most effective methods to obtain such data (for example, methods to choose optimal locations for new data collection).
- Understand whether models used for implementing regulation would be used widely by state and local governments and consulting firms that help those entities to develop implementation plans.

Modular Approaches for Environmental Model Development

The code of environmental models often can be written in a modular form. A module is an independent piece of code that forms a part of one or more models. Often, each module describes one process. For example, CMAQ includes modules for the description of horizontal and vertical advection, horizontal and vertical dispersion, gas-phase chemistry, aqueous-phase chemistry, aerosol thermodynamics and dynamics, plume chemistry effects, dry and wet deposition, and process analysis. This modular approach facilitates testing of the model (one can test the individual pieces separately) and reuse of the relevant modules in separate modeling projects. The parts of the model can be replaced with others without changing the overall structure of the model. There are also choices of modules for the same task. CMAQ allows its user to choose among three gas-phase chemistry mechanisms, depending on the specifics of the problem being modeled. The modular approach to CMAQ allows the level of complexity in the application to be aligned with the needs of the regulatory decisions. For example, the use of the mercury chemistry simulation capability of CMAQ is not necessary for ozone or particulate matter applications.

A major advantage of the modular model development approach is the ability to easily add or remove parts of the model, thus creating models of different complexity. For example, the full range of available modules (describing all potentially relevant processes) can be used, and then after quantifying the importance of each one of them for the specific application, the model can be simplified and used by removing the parts that have little or no effect on the results. For example, one can remove the cloud chemistry module from an application focusing on ozone episodes. The rest of the analysis can be done with the simplified model. Therefore, the modular approach and the resulting models of different degrees of complexity allow the user to satisfy the scientific requirements about quantifying the influence of the different processes and to avoid unnecessary complexity in the model used for the regulation. This approach allows modelers to defend their choices of excluding parts of the system from the analysis by allowing modelers to demonstrate the impacts of including or excluding various processes from the model. This approach also allows models to be updated more easily.

RECOMMENDATIONS

The committee offers several recommendations based on the discussion in this chapter. They deal with the interdependence of models and measurements, the model extrapolation, and the need for model parsimony.

The Interdependence of Models and Measurements

The interdependence of models and measurements is complex and iterative for several reasons. Measurements help to provide the conceptual basis of a model and inform model development, including parameter estimation. Measurements are also a critical tool for corroborating model results. Once developed, models can drive priorities for measurements that ultimately get used in modifying existing models or in developing new ones.

Measurement and model activities are often conducted in isolation. For example, modelers often add details to models without sufficient measurements to justify or confirm the importance of these changes. Likewise, field and laboratory scientists might expand their compilation of samples without understanding the utility of such information for modeling. Although environmental data systems serve a range of purposes, including compliance assessment, monitoring of trends in indicators, and basic research performance, the importance of models in the regulatory process requires measurements and models to be better integrated. Adaptive strategies that rely on iterations of measurements and modeling, such as those discussed in the 2003 NRC report titled *Adaptive Monitoring and Assessment for the Comprehensive Everglades Restoration Plan,* provide examples of how improved coordination might be achieved.

Recommendations

Using adaptive strategies to coordinate data collection and modeling should be a priority of decision makers and those responsible for regulatory model development and application. The interdependence of measurements and modeling needs to be fully considered as early as the conceptual model development phase. Developing adaptive strategies will benefit from the contributions of modelers, measurement experts, decision makers, and resource managers.

Model Parsimony

Models are always incomplete, and efforts to make them more complete can be problematic. As features and capabilities are added to a model, the cumulative effect on model performance needs to be evaluated carefully. Increasing the complexity of models without adequate consideration can introduce more model parameters with uncertain values, and decrease the potential for a model to be transparent and accessible to users and reviewers. It is often preferable to omit capabilities that do not improve model performance substantially. Even more problematic are models that accrue substantial uncertainties because they contain more parameters than can be estimated or calibrated with available observations.

Recommendations

Models used in the regulatory process should be no more complicated than is necessary to inform regulatory decisions. In the process of evaluating whether a model is suitable for its given application, there should be a critical evaluation of whether the model has been made unreasonably complicated. This evaluation should include how model developers and those that select a model for a particular application have addressed the trade-offs between the need for a given model application to be an accurate representation of the system of interest and the need for it to be reproducible, transparent, and useful for the regulatory decision at hand.

4

Model Evaluation

INTRODUCTION

How does one judge whether a model or a set of models and their results are adequate for supporting regulatory decision making? The essence of the problem is whether the behavior of a model matches the behavior of the (real) system sufficiently for the regulatory context. This issue has long been a matter of great interest, marked by many papers over the past several decades, but especially and distinctively by Caswell (1976) who observed that models are *objects* designed to fulfill clearly expressed tasks, just as hammers, screwdrivers, and other tools have been designed to serve identified or stated purposes. Although "model validation" became a common term for judging model performance, it has been argued persuasively (e.g., Oreskes et al. 1994) that complex computational models can never be truly validated, only "invalidated." The contemporary phrase for what one seeks to achieve in resolving model performance with observation is "evaluation" (Oreskes 1998). Although it might seem strange for such a label to be important, earlier terms used for describing the process of judging model performance have provoked rather vigorous debate, during which the word "validation" was first to be replaced by "history matching" (Konikow and Bredehoeft 1992) and later by the term "quality assurance" (Beck et al. 1997; Beck and Chen 2000). Some of these terms imply, innately or by their de facto use, a one-time approval step. Evaluation emerged from this debate as the most appropriate descriptor and is characteristic of a life-cycle process.

Two decades ago, model "validation" (as it was referred to then) was defined as the assessment of a model's predictive performance against a second set of (independent) field data given model parameter (coefficient) values identified or calibrated from a first set of data. In this restricted sense, "validation" is still a part of the common vocabulary of model builders.

The difficulty in finding a label for the process of judging whether a model is adequate and reliable for its task is described as follows. The terms "validation" and "assurance" prejudice expectations of the outcome of the procedure toward only the positive—the model *is* valid or its quality *is* assured—whereas evaluation is neutral in what might be expected of the outcome. Because awareness of environmental regulatory models has become so widespread in a more scientifically aware audience of stakeholders and the public, words used within the scientific enterprise can have meanings that are misleading in contexts outside the confines of the laboratory world. The public knows well that supposedly authoritative scientists can have diametrically opposed views on the benefits of proposed measures to protect the environment.

When there is great uncertainty surrounding the science base of an issue, groups of stakeholders within society can take this issue as a license to assert utter confidence in their respective versions of the science, each of which contradicts those of the other groups. Great uncertainty can lead paradoxically to a situation of "contradictory certainties" (Thompson et al. 1986), or at least to a plurality of legitimate perspectives on the given issue, with each such perspective buttressed by a model proclaimed to be *valid*. Those developing models have found this situation disquieting (Bredehoeft and Konikow 1993) because, even though science thrives on the competition of ideas, when two different models yield clearly contradictory results, as a matter of logic, they cannot both be true. It matters greatly how science and society communicate with each other (Nowotny et al. 2001); hence, in part, scientists shunned the word "validation" in judging model performance.

Today, evaluation comprises more than merely a test of whether history has been matched. Evaluation should not be something of an afterthought but, indeed, a process encompassing the entire life cycle of the task. Furthermore, for models used in environmental regulatory activities, the model builder is not the only archetypal interested party holding a stake in the process but is also one among several key players, including the model user, the decision maker or regulator, the regulated parties, and the affected members of the general public or the representative of the nongovernmental organization. Evaluation, in short, is an altogether much broader, more comprehensive affair than validation and encompasses more elements than simply the matching of observations to results.

This is not merely a question of form, however. In this chapter, where the committee describes the process of model evaluation, it adopts the perspective, discussed in Chapter 1 of this report, that a model is a "tool" designed to fulfill a task—providing scientific and technical support in the regulatory decision-making process—not a "truth-generating machine" (Janssen and Rotmans 1995; Beck et al. 1997). Furthermore, in sympathy with the *Zeitgeist* of contemporary environmental policy making, where the style of decision making has moved from that of a command-and-control technocracy to something of a more participatory, more open democracy (Darier et al. 1999), we must address the changing perception of what it takes to trust a model. This not only involves the elements of model evaluation but also who will have a legitimate right to say whether they can trust the model and the decisions emanating from its application. Achieving trust in the model among those stakeholders in the regulatory process is an objective to be pursued throughout the life of a model, from concept to application.

The committee's goal in this chapter is to articulate the process of model evaluation used to inform regulation and policy making. We cover three key issues: the essential objectives for model evaluation; the elements of model evaluation, and the management and documentation of the evaluation process. To discuss the elements of model evaluation in more detail, we characterize the life stages of a model and the application of the elements of model evaluation at these different stages. We organized the discussion around four stages in the life cycle of a regulatory model—problem identification, conceptual model development, model construction, and model application (see Figure 4-1). The life-cycle concept broadens the view of what modeling entails and may strengthen the confidence that users have in models. Although this perspective is somewhat novel, the committee observed some existing and informative examples in which model evaluations effectively tracked the life cycle of a model. These examples are discussed later in this chapter. We recognize that reducing a model's life cycle to four stages is a simplified view, especially for models with long lives that go through

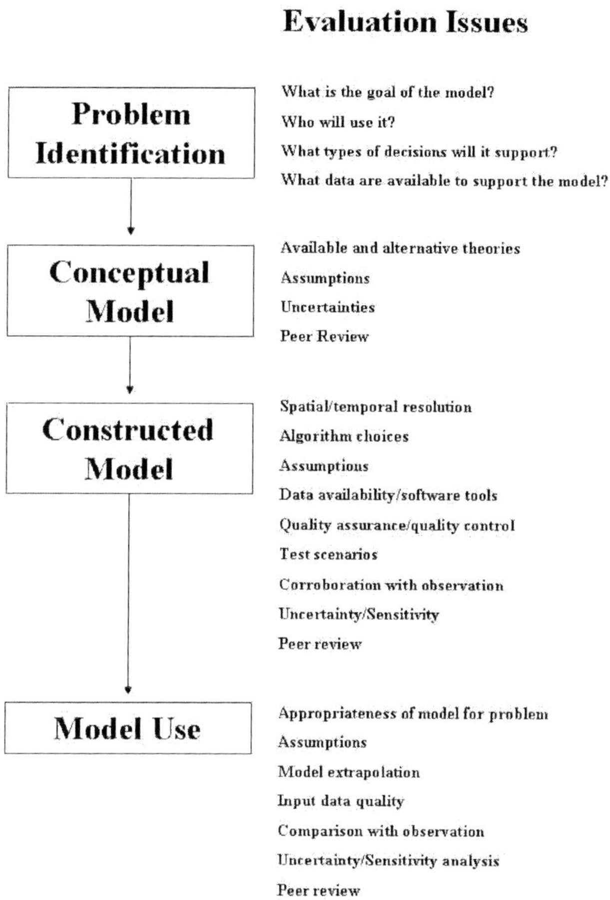

FIGURE 4-1 Stages of a model's life cycle.

important changes from version to version. The MOBILE model for estimating atmospheric vehicle emissions, the UAM (urban airshed model) air quality model, and the QUAL2 water quality models are examples of models that have had multiple versions and major scientific modifications and extensions in over two decades of their existence (Scheffe and Morris 1993; Barnwell et al. 2004; EPA 1999c). The perspective of a four-stage life cycle is also simplified from the stages of model development discussed in Chapter 3. However, simplifying a model's life cycle makes discussion of model evaluation more tractable.

Historically, the management of model quality has been inconsistent, due in part to the failure to recognize the impact of errors and omissions in the early stages of the life cycle of the model. At EPA (and other organizations), the model evaluation process traditionally has only begun at the model construction and model application stages. Yet formulating the wrong model questions or even confronting the right questions with the wrong conceptual model will result in serious quality problems in the use of a model. Limited empirical evidence in the groundwater modeling field suggests that 20-30% of model analyses confront new data that render the prevailing conceptual model invalid (Bredehoeft 2005). Such quality issues are difficult to discover and even more difficult to resolve (if discovered) when model evaluation applies only at the late stages of the model life cycle.

ESSENTIAL OBJECTIVES FOR MODEL EVALUATION

Fundamental Questions To Be Addressed

In the transformation from simple "validation" to the more extensive process of model evaluation, it is important to identify the questions that are confronted in model evaluation. When viewing model evaluation as an ongoing process, several key questions emerge. Beck (2002b) suggests the following formulation:

- Is the model based on generally accepted science and computational methods?
- Does it work, that is, does it fulfill its designated task or serve its intended purpose?
- Does its behavior approximate that observed in the system being modeled?

Responses to such questions will emerge and develop at various stages of model development and application, from the task description through the construction of the conceptual and computational models and eventually to the applications. The committee believes that answering these questions requires careful assessment of information obtained at each stage of a model's life cycle.

Striving for Parsimony and Transparency

In the development and use of models, parsimony refers to the preference for the least complicated explanation for an observation. Transparency refers to the need for stakeholders and members of the public to comprehend the essential workings of the model and its outputs. Parsimony derives from Occam's (or Ockham's) razor attributed to the 14th century logician William of Occam, stating that "entities should not be multiplied unnecessarily." Parsimony does not justify simplicity for its own sake. It instead demands that a model capture all essential processes for the system under consideration—but no more. It requires that models meet the difficult goal of being accurate representations of the system of interest while being reproducible, transparent, and useful for the regulatory decision at hand.

The need to move beyond simple validation exercises to a more extensive model evaluation leads to the need for EPA to explicitly assess the trade-offs that affect parsimony, transparency, and other considerations in the process of developing and applying models. These trade-offs are important to modelers, regulators, and stakeholders. The committee has identified three fundamental goals to be considered in making trade-offs, which are further discussed in Box 4-1:

- *The need to get the correct answer* – This goal refers to the need to make a model capable of generating accurate as well as consistent and reproducible projections of future behavior or consistent assessments of current relationships.
- *The need to get the correct answer for the correct reason* – This goal refers to the reproduction of the spatial and temporal detail of what scientists consider to be the essence of the system's workings. Simple process and empirical models can be "trained" to mimic a system of interest for an initial set of observations, but if the model fails to capture all the important system processes, the model could fail to behave correctly for an observation outside the limited range of "training" observations. Such failure tends to drive models to be more detailed.
- *Transparency* – This goal refers to the comprehension of the essential workings of the model by peer reviewers as well as informed but scientifically lay stakeholders and members of the public. This need drives models to be less detailed. Transparency can also been enhanced

BOX 4-1 Attributes That Foster Accuracy, Precision, Parsimony, and Transparency in Models

Gets the Correct Result

- Model behavior closely approximates behavior of real system
  - High predictive power on a case-by-case basis
  - High predictive power a statistical basis
- Model results insensitive to factors that should not affect them

Gets the Correct Result for the Right Reason

- Model accurately represents the real system
  - Comprehensive
    - Variables
      - Inputs, outputs
      - Exogenous, endogenous
    - Relationships
      - Functional
      - Cause-effect
    - Statistical Circumstances
      - Input changes
      - Assumption relaxation
    - Resolutions
      - Temporal
      - Spatial
- Model is based on good science
  - Accepted principles, theory, results
    - From peer reviewed sources
    - Prestige of developer or lab
  - Up-to-date
    - Concepts and theory
    - Algorithms, computational methods
    - Empirical findings
- Appropriate data are available or feasible to acquire
  - Estimates for model parameters
  - Data for model calibration

Transparency

- Suits specific regulatory context or decisions
  - Address the specific concern
  - Usable by
    - Decision makers
    - Implementers
  - Understandable by
    - Decision makers
    - Stakeholders
    - Implementers

- Model is seen to be appropriate for the specific system
  - Application is within model limitations
    - Resolution
    - Parameter values
    - Special system characteristics (for example, special weather characteristics or soil chemistry)
  - Inputs available for the specific system
    - Parameter estimates
    - Calibration data
- Results/outputs are helpful
  - Interpretable
  - Relate to regulatory objectives
    - Decision makers
    - Stakeholders
  - Are "actionable," i.e., they relate to decision variables or policy parameters understandable to decision makers, stakeholders, and the informed public

by ensuring that reviewers, stakeholders and the public comprehend the processes followed in developing, evaluating, and applying a model, even if they do not fully understand the basic science behind the models.

These three goals can result in competing objectives in model development and application. For example, if the primary task was to use a model as a repository of knowledge, its design might place priority on getting sufficient detail to ensure that the result is correct for the correct reasons. On the other hand, to meet the task of the model as a communication device, the optimal model would minimize detail to ensure transparency. It is also of interest to consider when a regulatory task would be best served by having a model err on the side of getting accurate results but not including sufficient detail to match scientific understanding. For example, when an exposure model can accurately define the relationship between a chemical release to surface water based on a detailed mass balance, should the regulator consider an empirical model that has the same level of accuracy? Here, parsimony might give preference to the simpler empirical model, whereas transparency is best served by the mass-balance model that allows the model user to see how the release is transformed into a concentration. Moreover, in the regulatory context, the more-detailed model addresses the need to reveal to decision makers and stakeholders how different environmental processes can affect the link from emissions to concentration. Nevertheless, if the simpler empirical model provides both accurate and consistent results, it should have a role in the decision process even if that role is to provide complementary support and evaluation for the more-detailed model.

The committee finds that modelers may often err on the side of making models more detailed than necessary. The reasons for the increasing complexity are varied, but one regulatory modeler mentioned that it is not only modelers that strive to building a more complex model but also stakeholders who wish to ensure that their issue or concerns are represented in the model, even if addressing such concerns does not have an impact on model results (A. Gilliland, Model Evaluation and Applications Branch, Office of Research and Development, EPA, personal commun., May 19, 2006). Increasing the refinement of models introduces increasing model parameters with uncertain values while decreasing the model transparency to users and reviewers. Here, the problem is a model that accrues significant uncertainties when it contains more parameters than can be calibrated with observations available to the model evaluation process. In spite of the drive to make their models more detailed, modelers often prefer to omit capabilities that do not substantially improve model performance—that is, its precision and accuracy for addressing a specific regulatory question.

ELEMENTS OF MODEL EVALUATION

The evidence used to judge the adequacy of a model for decision-making purposes comes from a variety of sources. They include studies that compare model results with known test cases or observations, comments from the peer review process, and the list of a model's major assumptions. Box 4-2 lists those and other elements of model evaluation. Many of the elements might be repeated, eliminated, or added to the evaluation as a model's life cycle moves from problem identification to model application stages. For example, peer review at the model development stage might focus on the translation of theory into mathematical algorithms and numerical solutions, whereas peer review at the model application stage might focus on the adequacy of the input parameters, model execution, and stakeholder involvement. Recognizing that model evaluation may occur separately during the early stages of a model's life, as well as again during subsequent applications, helps to address issues that might arise when a model is applied by different groups and for different conditions than those for which the model was developed. The committee notes that, whereas the elements of model evaluation and the questions to be answered throughout the evaluation process may be generic in nature, what comprises a high-quality evaluation of a model will be both task- and case-specific. As described in Chapter 2, the use of models in environmental regulatory activities varies widely both in the effort and the consequences of the regulatory efforts it supports. Thus, the model evaluation process and the resources devoted to it must be tailored to its specific context. Depending on the setting, model evaluation will not necessarily address all the elements listed in Box 4-2. In its guidance document on the use of models at the agency, EPA (2003d) recognized that a model evaluation should adopt a graded approach to model evaluation, reflecting the need for it to be adequate and appropriate for the decision at hand. The EPA Science Advisory Board (SAB) in its review of EPA's guidance document on the use of models recommended that the graded concept be expanded to include model development and application (EPA 2006d). The committee here recognizes that model evaluation must be tailored to the complexity and impacts at hand as well as the life stage of the model and the model's evaluation history.

MODEL EVALUATION AT THE PROBLEM IDENTIFICATION STAGE

There are many reasons why regulatory activities can be supported by environmental modeling. At the problem identification stage, decision makers together with model developers and other analysts must consider the regulatory decision at hand, the type of input the decision needs, and whether and how modeling can contribute to the decision-making process. For example, if a regulatory problem involves the assessment of the health risk of a chemical, considerations may include whether to focus narrowly on cancer risk or to include a broader spectrum of health risks. Another consideration might be whether the regulatory problem focuses on occupational exposures, acute exposures, chronic exposures, or exposures that occur to a susceptible subpopulation. The final consideration is whether a model might aid in the regulatory activity.

If there is sufficient need for computational modeling, there are three questions that must be addressed at the problem identification stage: (1) What types of decisions will the model support? (2) Who will use it? and (3) What data are available to support development, application, and evaluation of a model? Addressing these questions is important

BOX 4-2 Individual Elements of Model Evaluation

Scientific basis – The scientific theories that form the basis for models.

Computational infrastructure – The mathematical algorithms and approaches used in the execution of the model computations.

Assumptions and limitations – The detailing of important assumptions used in the development or application of a computational model as well as the resulting limitations in the model that will affect the model's applicability.

Peer review – The documented critical review of a model or its application conducted by qualified individuals who are independent of those who performed the work, but who are collectively at least equivalent in technical expertise (i.e., peers) to those who performed the original work. Peer review attempts to ensure that the model is technically adequate, competently performed, properly documented, and satisfies established quality requirements through the review of assumptions, calculations, extrapolations, alternate interpretations, methodology, acceptance criteria, and/or conclusions pertaining from a model or its application (modified from EPA 2006a).

Quality assurance and quality control (QA/QC) – A system of management activities involving planning, implementation, documentation, assessment reporting, and improvement to ensure that a model and its component parts are of the type needed and expected for its task and that they meet all required performance standards.

Data availability and quality – The availability and quality of monitoring and laboratory data that can be used for both developing model input parameters and assessing model results.

Test cases – Basic model runs where an analytical solution is available or an empirical solution is known with a high degree of confidence to ensure that algorithms and computational processes are implemented correctly.

Corroboration of model results with observations – Comparison of model results with data collected in the field or laboratory to assess the accuracy and improve the performance of the model.

Benchmarking against other models – Comparison of model results with other similar models.

Sensitivity and uncertainty analysis – Investigation of what parameters or processes are driving model results as well as the effects of lack of knowledge and other potential sources of error in the model.

Model resolution capabilities – The level of disaggregation of processes and results in the model compared to the resolution needs from the problem statement or model application. The resolution includes the level of spatial, temporal, demographic or other types of disaggregation.

Transparency – The need for individuals and groups outside modeling activities to comprehend either the processes followed in evaluation or the essential workings of the model and its outputs.

both for setting the direction of the model and for setting goals for the quality and quantity of information needed to construct and apply the model.

At this stage, data considerations should be a secondary issue, though not one to completely ignore. Problem identification must not be anchored solely to the available data to avoid the situation where data dictate the problem identification of the form, "We have these data available, so we can answer this question. . . ." However, there would have to be confidence that quantitative analysis could inform the problem and that some data would be available.

The problem identification stage answers the question of whether modeling might help to inform the particular issue at hand and sets the direction for development of conceptual and computation models. Although the committee is not endorsing a complex model evaluation at the nascent stage of problem identification, it is clear that setting off to develop or apply a model that will not address the problem at hand or that will take too long to provide answers can have serious impacts on the effectiveness of modeling. The key goal of the problem identification phase is to identify the regulatory task at hand and assess the role that modeling could play. At this stage, the description of the regulatory task and the way modeling might address this regulatory task should be open to comment and criticism. Thus, when formal model evaluation is performed in later stages of a model's life cycle, it must take into account the problem identification and how it influenced the nature of the model.

EVALUATION AT THE CONCEPTUAL MODEL STAGE

Some of the most important model choices are made at the conceptual stage, yet most model evaluation activities tend to avoid a critical evaluation at this stage. Often a peer review panel will begin its efforts with the implicit acceptance of all the key assumptions made to establish the conceptual model and then devote all of its attention to the model building and model application stages. Alternatively, a late-stage peer review of a nearly complete model may find the underlying conceptual model to be flawed. Finally, data must be assessed at this point to ensure the availability of data for model development, input parameters, and evaluation. The result of this process is the selection of a computational modeling approach that addresses problem identification, data availability, and transparency requirements.

Evaluating the Conceptual Model

Quality of the Basic Science

It is important to evaluate the fundamental science that forms the basis of the conceptual model. One approach is to consider the idea of a *pedigree* of a domain of science, a word expressing something about the history—and the quality of the history—of the concepts and theories behind the model and, possibly more appropriately, each of its constituent parts (Funtowicz and Ravetz 1990). Over the years, the fundamental scientific understanding and other understandings that are used in constructing models have been consolidated and refined to produce a mature product with a pedigree. For example, a task, such as modeling of lake eutrophication, started as an embryonic field of study, passed through the adolescence of competing schools of thought (Vollenweider 1968) to the gathering of consensus around a single scientific outlook (disputed only by the sub-discipline's "rebels"), and finally to the adulthood of the fully consolidated outlook, contested, if at all, only by those considered "cranks" by the overwhelming majority—a history partially recounted in Schertzer and Lam (2002). The status of a model's pedigree typically changes over time, with the strong implication of ever-improving quality. Although some models may cease to improve over time, it is more common that they continue to be refined over time, especially for long-lived regulatory models. The concept of a pedigree can be applied to the model as a whole, to one of its major subblocks (such as atmospheric chemistry or human toxicology), or to each constituent model parameter.

Quality of Available Data

For environmental models, one of the issues often ignored at the conceptual stage is the availability of data. It is one of the major issues in the use of environmental models, and it has multiple aspects:

- Data used as inputs to the model, including data used to develop the model.
- Data used to estimate values of model parameters (internal variables).
- Data used for model evaluation.

There is some overlap between the first and second types of data, depending on the model application, but in general these data needs can be viewed as separate. One major problem is that collecting new data at this early stage is rarely considered. Model development and evaluation and data collection should be iterative and proceed together, but in practice, these activities at agencies such as EPA often are done by separate groups that may not meet each other until late in the process. The critical issue is that, at this stage in a model's life cycle, there should be a requirement for an assessment of the data needs and a corresponding data collection plan. Modelers should be building on-going collaborations with experimentalists and those responsible for collection of additional data to determine how such new data can guide model development and how the resulting models can guide the collection of additional data.

EVALUATION AT THE COMPUTATIONAL MODEL STAGE

In moving from the identification of the problem, the assessment of required resolving power and data needs, and the decision concerning the basic qualitative modeling approach to a constructed computational model, a number of practical considerations arise. As we observe in Chapter 3, these considerations include (1) choices of specific mathematical expressions to represent the interactions among the model's state variables; (2) evaluation of a host of algorithmic and software issues relating to numerical solution of the model's equations; (3) the assembly of data to develop inputs, to test, and to compare with model results; and (4) the ability of the model to arrange the resulting numerical outputs for comprehension by all the stakeholders concerned. A prime motivation at this stage of evaluation is, does the behavior of the model approximate well what is observed? For modelers, nothing is more convincing and reassuring than seeing the curve of the model's simulated responses passing through the dots of observed past behavior. However, as discussed in Chapter 1, natural systems are never closed and model results are never unique. Thus, any match between observations and model results might be due to processes not represented in the model canceling each other out. In addition, simply reproducing results that match observations for a single scenario or several scenarios does not mean the model can represent the full statistical characteristics of observations.

The evaluation needs fundamentally to address the questions laid out at the beginning of this chapter: the degree to which the model is based on generally accepted science and computational methods, whether the model fulfills its designed task, and how well its behavior approximates that observed in the system being modeled. A majority of model evaluation activities traditionally occur at the stages in which the computational model is developed and applied. These are the stages when quality assurance and quality control (QA/QC) efforts are documented, testing and analysis reports generated, model documentation produced, and peer review panels commissioned. However, these formal model evaluation activities must be cognizant of and built on earlier evaluation activities during the problem identification and model conceptualization stages.

Scientific Basis, Computational Infrastructure, and Assumptions

The scientific basis, the computational infrastructure, and the major assumptions used within a computational model are some of the first elements typically addressed during model evaluation. The initial evaluation of the scientific theories, possible computational approaches, and inherent assumptions should occur during the development of the conceptual model. Model builders must reassess these issues during the construction of a computational model by obtaining a wider array of peer reviewers' and others' comments. Indeed, these issues are typically the first elements assessed by outside evaluators when EPA models go before review panels, such as the SAB, or the public.

Code Verification of Numerical Solutions and Other Quality Assurance Procedures

Verification of model code and assurance that the numerical algorithms are operating correctly are the essence of QA/QC procedures. These activities evaluate to what extent the executable code and other numerical software in the constructed model generate reliable and consistent solutions to the mathematical equations of the model. The document prepared for a recent evaluation by SAB of the very-high-order 3MRA modeling system (the multimedia model described in Babendreier and Castleton [2005]) defines code verification as follows (EPA 2003e):

> *Verification* refers to activities that are designed to confirm that the mathematical framework embodied in the module is correct and that the computer code for a module is operating according to its intended design so that the results obtained using the code compare favorably with those obtained using known analytical solutions or numerical solutions from simulators based on similar or identical mathematical frameworks.

Verification activities include taking steps to ensure that code is properly maintained, modified to correct errors, and tested across all aspects of the module's functionality. Table 4-1 lists some of the software checks listed by EPA to ensure that model computations proceed as anticipated. Other QA/QC activities include (1) the use of the model in different operating systems with different compilers to make sure that the results remain the same and (2) testing under simplified scenarios (for example, with zero emissions, zero boundary conditions, and zero initial conditions) where an analytical solution is available or an empirical solution is known with a high degree of confidence.

Like so many things, concluding—provisionally—that the constructed model is working with a reliable code comes down to the outcomes of the most rudimentary tests, such as those "comparing module results with those generated independently from hand calculations or spreadsheet models" (EPA 2003e). These tests are the equivalent of the tests made time and again to ensure a sensor or instrument is working properly. They are tests that are maximally robust against ambiguous outcomes. As such, they only ensure against gross deficiencies but cannot confirm that a model is sufficiently sound for regulatory use. Constant vigilance is required. "Even legacy codes that had more than a decade of wide use experienced environmental conditions that caused unstable numerical solutions" (EPA 2003e).

Where models are linked, as in linking emissions models to fate and transport models as discussed in Chapter 2, additional checks and audits are required to ensure the streams of data passing back and forth have strictly identical meanings and units in the partnered codes engaging in these electronic transactions. Further, such linked models do not lend themselves to be compared with simple test cases that have known solutions. This makes QA/QC activities related to linked models much more difficult.

TABLE 4-1 QA/QC Checks for Model Code

Software code development inspections: Software requirements, software design, or code are examined by an independent person or groups other than the author(s) to detect faults, programming errors, violations of development standards, or other problems. All errors found are recorded at the time of inspection, with later verification that all errors found have been successfully corrected.

Software code performance testing: Software used to compute model predictions is tested to assess its performance relative to specific response times, computer processing usage, run time, convergence to solutions, stability of the solution algorithms, the absence of terminal failures, and other quantitative aspects of computer operation.

Tests for individual model module: Checks ensure that the computer code for each module is computing module outputs accurately and within any specific time constraints. (Modules are different segments or portions of the model linked together to obtain the final model prediction.)

Model framework testing: The full model framework is tested as the ultimate level of integration testing to verify that all project-specific requirements have been implemented as intended.

Integration tests: The computational and transfer interfaces between modules need to allow an accurate transfer of information from one module to the next, and ensure that uncertainties in one module are not lost or changed when that information is transferred to the next module. These tests detect unanticipated interactions between modules and track down cause(s) of those interactions. (Integration tests should be designed and applied in a hierarchical way by increasing, as testing proceeds, the number of modules tested and the subsystem complexity.)

Regression tests: All testing performed on the original version of the module or linked modules is repeated to detect new "bugs" introduced by changes made in the code to correct a model.

Stress testing (of complex models): Stress testing ensures that the maximum load (for example, real time data acquisition and control systems) does not exceed limits. The stress test should attempt to simulate the maximum input, output, and computational load expected during peak usage. The load can be defined quantitatively using criteria such as the frequency of inputs and outputs or the number of computations or disk accesses per unit of time.

Acceptance testing: Certain contractually required testing may be needed before the new model or model application is accepted by the client. Specific procedures and the criteria for passing the acceptance test are listed before the testing is conducted. A stress test and a thorough evaluation of the user interface is a recommended part of the acceptance test.

Beta testing of the pre-release hardware/software: Persons outside the project group use the software as they would in normal operation and record any anomalies encountered or answer questions provided in a testing protocol by the regulatory program. The users report these observations to the regulatory program or specified developers, who address the problems before release of the final version.

Reasonableness checks: These checks involve items like order-of-magnitude, unit, and other checks to ensure that the numbers are in the range of what is expected.

Source: EPA 2002e.

Comparing Model Output to Data

Comparing model results with observations is a central component of any effort to evaluate models. However, such comparisons must be made in light of the model's purpose—a tool for assessment or prediction in support of making a decision or formulating policy. The inherent problems of providing an adequate set of observations and making credible comparisons give rise to some important issues.

The Role of Statistics

Because (near) perfect agreement between model output and observations cannot be expected, statistical concepts and methods play an inevitable and essential role in model evaluation. Indeed, it is tempting to use formal statistical hypothesis testing as an evaluation tool, perhaps in part because such terms as "accepting" and "rejecting" hypotheses sound as though they might provide a way to validate models in the now-discredited meaning of the term. However, the committee has concerns that testing (for example, that the mean of the observations equals the mean of the model output) will fail to provide much insight into the appropriateness of using an environmental model in a specific application. As discussed in Box 1-1 in Chapter 1, the evaluation of the ozone models in the 1980s and early 1990s showed that estimates of ozone concentrations from air quality models were good when compared with observations for any choice of statistical methods, but only because the errors in the models tended to cancel out. Statistics has value for conceptualizing, visualizing, and quantifying variation and dependence rather than for serving as a source of "rigorous" or "objective" standards for model evaluation. The committee cautions, however, that standard, elementary statistical methods will often be inappropriate in environmental applications, for which problems of spatial and temporal dependence are frequently a critical issue.

Although epidemiologists and air quality modelers use statistical tests to compare models with data, it is difficult to find broad-based examples in regulatory models in which formal hypothesis testing (e.g., testing that the means of two distributions are equal based on the p-value of some test statistic) has played a substantial role in any model evaluation. What is needed is statistically-sophisticated analysts that can do non-standard statistical analyses appropriate for the individual circumstances. For example, air quality modelers commonly present a variety of model performance statistics along with graphic comparisons of model results with observations; these are sometimes compared with acceptability criteria set by EPA for various applications.

Comparing Models with Data—Model Calibration

Model calibration is the process of changing values of model input parameters within a physically defensible range in an attempt to match the model to field observations within some acceptable criteria. Models often have a number of parameters whose values cannot be established in the model development stage but must be "calibrated" during initial model testing. This need requires observations for conditions that must broadly characterize the conditions for which the model will be used. Lack of characterization of the conditions can result in a model that is calibrated to a set of conditions that are not representative of the range of conditions for which the model is intended. The calibration step can be linked with a "validation" step where a portion of the observations are used to calibrate the model, and then the calibrated model is run and results compared with the other portion of data to "validate" the model. The typical criteria used for judging the quality of agreement is mean square error, or the average squared difference between observed values and the values predicted by the model.

The issue of model calibration can be contentious. The calibration tradition is ingrained in the water resources field by groundwater, stream-flow, and water-quality modelers, whereas the practice is shunned by air-quality modelers. This practice is not merely a disagreement about terminology, but a more fundamental difference of opinion about the relationship of models and measurement data, which is explored in Box 4-3. However, it is clear that both fields, and modelers in general, accept a fundamental role for measurement data to improve modeling. In this unifying view, model calibration is not just a matter of fiddling about trying to find suitable best values of the coefficients (parameters) in the model. Instead, calibration has to do with evaluating and quantifying the posterior uncertainty attached to the model as a function of the measured data, prior model uncertainty, and the uncertainty in the measured data against which it has been reconciled (calibrated). This view is clearly Bayesian in spirit, using data and prior knowledge to arrive at updated posterior expectations about a phenomenon, if not strictly so in number-crunching, computational terms. It is the recognition of the fundamental codependence of models and data from measurements that is common among all models.

BOX 4-3 To Calibrate or Not To Calibrate

In an ideal world, calibration of models would not be necessary, at least not if we view calibration merely as the search for values of a model's parameters that provide the best match of the model's behavior with that observed of the real system. It would not be necessary because the model would contain only parameters that are known to a high degree of accuracy. To be more pragmatic, but nonetheless somewhat philosophical, there is a debate about whether to calibrate a model or not in the real world of environmental modeling. That debate centers around two features: (1) the principle of engaging models with field data in a learning context during the development of the model; and (2) the principle of using calibration for quantifying the levels of uncertainty attached to the model's parameters, with a view to accounting for how that uncertainty propagates forwards with predictions. The former lies within the conventional understanding and interpretation of what constitutes model calibration. The latter requires a broader, but less familiar, interpretation of calibration. Taken together, calibration can be seen to be something more than a "fiddler's paradise," in which the analyst seeks merely to fit the data, no matter how absurd the resulting parameter estimates; and no matter the obvious risk of subsequently making confident—but probably highly erroneous—predictions of future behavior, especially under conditions different from those reflected in the data used for model calibration (Beck 1987).

The nub of the debate turns on the extent to which the analyst trusts the prior knowledge about the *individual* components of the model, to which the parameters are attached, yet discounts the power of the calibration data set—reflecting the collective effects of all the model's parameters on observed behavior of the prototype, as a *whole*—to overturn these presumptions. The debate also turns on the extent to which individual parameters can be "measured" independently in the field or laboratory under tightly controlled conditions. The more this is feasible, the less the need to calibrate the behavior of the model (as a whole). In this argument, however, it must be remembered that many parameters remain quantities that appear in presumed relationships, that is, mathematical relationships or models between the observed quantities, so that the problem of calibrating the model as a whole is transferred to calibrating the relationship between the observables to which each individual parameter is bound. This may seem less of a problem when needing to substitute a value for soil porosity into a hydrological model. But it is surely a problem when the need is to find a value for a maximal specific growth-rate constant for a bacterial population, which is certainly not a quantity that can itself be directly measured.

Experience of model calibration and the stances taken on it differ from one discipline to another. In hydrology and water quality modeling it is unsurprising how the wider interpretation and greater use of calibration have become established practice. In spite of the relatively large volumes of hydrological field data customarily available, experience over several decades has shown that hydrological and water quality models inevitably suffer from a lack of identifiability in that many combinations of parameter values will enable the model to match the data reasonably well (Jakeman and Hornberger 1993; Beven 1996). Trying to find a best set of parameter values for the model, even a best structure for the model, have come to be accepted as barely achievable goals at best. In a pragmatic, decision-support context, what matters—given uncertain models, uncertain data, and therefore uncertain model forecasts—is whether any particular course of action (among the various options) manages to stand out above the fog of uncertainty as clearly the preferred option. Under this view, the posterior parametric uncertainties reflect the signature, or fingerprint, of all the distortions and uncertainties remaining in the model as a result of reconciling it with the field data. In a more theoretical context, interpretation of the patterns of such distortions and uncertainties can serve the purpose of learning from having engaged the model systematically with the field data.

In other disciplines, such as modeling of air quality, calibration is viewed as a practice that should be avoided at all costs. Inputs to these models include pollutant emissions (spatially, temporally, and chemically resolved), three-dimensional meteorological fields (such as wind speed and direction, temperature, relative humidity, sunlight intensity, clouds and rain, also temporally resolved). Air quality models also rely on a wide range of parameters used in the description of processes simulated by the models (such as turbulent dispersion coefficients for atmospheric mixing, parameters for the dry and wet removal of pollutants, kinetic coefficients for gas and aqueous-phase chemistry, mass transfer rate constants, and thermodynamic data for the partitioning of pollutants among the different phases present in the atmosphere).

The need for the determination of all of these input values and parameters has resulted in a huge investment in scientific research funded by EPA, state air pollution authorities (especially California), National Science Foundation (NSF), and others to understand the corresponding processes and to develop model application-independent approaches to estimate them. Further, complex regional meteorological models (such as MM5 and RAMS), which are used for other applications, are used to simulate the meteorology of the atmosphere and provide the corresponding input fields to the air quality models. Meteorological models themselves take advantage of the available measurements of wind speed, temperature, relative humidity, etc. in the domain that they simulate, to improve their predictions. In a technique called data assimilation the available measurements are used to "nudge" the meteorological model predictions closer to the available measurements by adding forcing terms (proportional to the difference between the model predictions and the observations) to the corresponding differential equations solved by the model. This semi-empirical form of correction can maintain the meteorological model results close to reality and improve the quality of the input provided to the air quality model. This form of calibration is involved only in the preparation of the input to the air quality model and is independent of the air quality model, its prediction, and the available air quality modeling.

The emission fields are prepared by corresponding emission models that incorporate the available information about the activity levels (for example, traffic, fuel consumption by industries, population density, etc.) and emission factors

(Continued on next page)

(emissions per unit of activity) for each source. Some of the best applications of air quality models have been accompanied by field measurements of emissions during the model application period (for example, transportation emissions in tunnels in the area, characterization of major local sources, even use of airplanes to characterize the plumes of major point sources, etc.). Boundary conditions are measured usually by ground monitoring stations or airplanes in selected points close to the model boundary (for example, in San Nicolas Island off the shore of Southern California). Laboratory (for example in smog chambers simulating the atmosphere) and field experiments have been used to understand the corresponding processes and to provide the necessary parameters.

One could argue that the historical lack of reliance on model calibration for the air quality area has resulted in significant research to understand better the most important processes and in the development of approaches to provide the necessary input. This has required a huge investment by US funding sources (the State of California, EPA, NSF, etc.) but has also resulted in probably the most comprehensive modeling tools available for environmental regulation. One could also argue that the atmosphere is a much easier medium to model (after all air is the same everywhere) compared to soil, water, ecosystems, or the human body. However, the success of the "let's try to avoid calibration" philosophy may be a good example in the long term for other environmental modeling areas.

In sum, there is nothing wrong with the healthy debate over calibration. Either way—whether calibration is accepted practice or shunned—all agree that fitting a model to past data is not an end in itself, but a means: to the end of learning something significant about the behavior of the real system; and to the end of faithfully reflecting the ineluctable uncertainty in a model.

One effect of the rejection of model calibration for regional air quality models is the idea that model results are more appropriate for relative comparisons than for absolute estimates. EPA guidance for the use of models for the attainment of ambient air quality standards (the attainment demonstration) for 8-hour ozone and the fine-particle particulate matter (PM) begins with the notion that model estimates will not predict perfectly the observed air quality at any given location at the present time and in the future (EPA 2005d). Thus, models for demonstrating whether emissions reduction strategies will result in attainment demonstrations are recommended for use in a relative sense in concert with observed air quality data. Such use essentially involves taking the ratio of future to present predicted air quality from the models to develop a ratio and then multiplying it by an "ambient" design value. The effect will be to anchor future concentrations to "real" ambient values. If air quality models were calibrated to observations, as is done with water quality models, there would be less need to use the model in a relative sense.

EPA also uses the concept that air quality models are imperfect predictors to argue for a weight-of-evidence approach to attainment demonstrations. Under a weight-of-evidence approach, the results of the air quality models are no longer the sole determining factor but rather one input that may include trends in ambient air quality and emissions observations and other information (EPA 2005d).

Comparing Models with Data—Data Quality

Not all data are of equal quality. In addition to the usual issues of systematic and random measurement errors, there is the issue that some "data" are the result of processing sensor information through instrumentation algorithms that are really models in their own right. Examples include the post-processing of raw information that is obtained from remote-sensing instruments (e.g., Henderson and Lewis 1998) or from techniques used to separate total carbon in an airborne PM sample into inorganic and organic carbon components (e.g., Chow et al. 2001). Thus, if the data and model output disagree, the extent of disagreement that is due to the model used to convert raw measurements into the quantity of interest must be considered. An additional and related difficulty with many data sets is that the standard assumption of statistically independent measurement errors can be untrue, including for remotely sensed data, greatly complicating model and measurement data comparisons.

Comparing Models with Data—Temporal and Spatial Issues

Even with data of impeccable quality, there are still many problems in comparing them with model output. One problem is that data and model output are generally averages over different temporal and spatial scales. For example, air pollution monitors produce an observation at a point, whereas output from regional-scaled air quality models discussed earlier in the report produces at best averages over the grid cells used in the numerical solution of the governing partial differential equations. However, if for no other reason than that the meteorological inputs into air pollution models will inevitably have errors at small spatial scales, there is no expectation that the models would reproduce actual average pollution levels over the grid cells, even if such an average could be observed. The models may do somewhat better at reproducing averages over larger regions of space or over longer intervals of time than the nominal observation frequency, and a model that does well with such averages could reasonably be judged as functioning well. Similar problems underlie many health assessments, such as when pharmacokinetic models for one exposure scenario are compared with measurements from a different exposure scenario or when data from laboratory rats exposed for 90 days are used to estimate human risks from a continuous lifetime exposure. Even so, these dilemmas are the reason models are needed—it is impossible to measure all events of interest.

There are two potential approaches that can address some of these spatial and temporal problems. The collection of two or more measurements inside the same computational cell provides information on the spatial variability of the pollutant of interest within a grid cell. However, monitoring is not always available to obtain multiple samples within the same grid cell. For the temporal issue, the collection of high temporal resolution measurements, including continuous measurements, can allow the comparison to be performed at several different time-intervals. In this manner, a model could be "stressed" to produce, for example, diurnal profiles of the pollutant. Again, however, the availability of monitoring data is a limiting factor.

Comparing Models with Data—Simulating Events Versus Long-Term Averages

An important issue is whether models are expected to reproduce observations on an event-by-event basis. If the model is used for short-term assessment or forecasting, then such a capability would be necessary. For example, when assessing whether an urban storm-water control system would be overwhelmed, resulting in the discharge of combined storm-water sewage into receiving waters, only a single-event rainfall-runoff model might be required to treat each potential storm event individually. However, when the goal is to predict how the environment will change over the long term in response to an EPA policy, such a capability is neither necessary nor sufficient. General circulation models used for assessing climatic change may be an extreme example of models that cannot reproduce event-by-event observations but are able to reproduce many of the statistical characteristics of climate over long-tern scales.

Comparing Models with Data—Simulating Novel Conditions

The comparison of model and measured data under existing conditions, no matter how extensive, provides only indirect evidence of how well a model will do at predicting what will happen under novel or post-regulatory conditions. Yet, this comparison is a fundamental element of model evaluation and its relevancy is perhaps the biggest challenge EPA faces in assessing the usefulness of its models. When model results are to be extrapolated outside of conditions for which they have been evaluated, it is important that they have the strongest possible theoretical basis, explicitly representing the processes that will most affect outcomes in the new conditions to be modeled, and embodying the best possible parameter estimates. For some models, such as for air dispersion models, it may be possible to compare output with data in a wide enough variety of circumstances to gain confidence that they will work well in new settings. Satisfying all of these conditions, however, is not always possible, as the case of competing cancer potency dose-response models makes clear. Absent a solid understanding of underlying mechanisms, the best model for doing such an extrapolation is a matter of debate.

There is the potential to test some types of models in cases where the system behaves differently, such as when there is a significant change in pollutant loads. Air pollution studies have indicated that air quality models can be stressed by simulating special periods, such as the Christmas holidays, with its low traffic emissions and high wood burning; days with major power disruptions (for example, the blackouts in the Northeast); or days when most people go on vacation (as in Europe). Pope (1989) provides an example of the possible insights from developing a model under such novel conditions. This study used epidemiological modeling to look at the reduction in hospital admissions for pneumonia, bronchitis, and asthma that occurred in the Utah Valley when a major source of pollution, the local steel mill, was closed for 13 months. The observation of a statistically significant reduction in hospital visits correlated to reductions in ambient PM concentrations helped to initiate a reassessment of ambient air quality standards for this pollutant.

Comparing Models with Data—A Bayesian Approach

For models that are used frequently, a Bayesian approach might be considered to quantitatively support model evaluation (Pascual 2004;

Reckhow 2005). For example, prior uses of the model could provide comparison of pre-implementation predictions of the success of an environmental management strategy with post-implementation observations. Using Bayesian analysis, this "prior" could be combined with a prediction-observation comparison for the site and topic of interest to evaluate the model as well as improve the strategy.

Uncertainty Analysis

Formal uncertainty analysis provides model developers, decision makers, and others with an assessment of the degree of confidence associated with model results as well as the aspects of the model having the largest impacts on its results. As such, uncertainty analysis and related sensitivity analysis is a critical aspect of model evaluation during model development and model application stages. The use of formal qualitative and quantitative uncertainty analysis in environmental regulatory modeling is growing in response to improvements in methods and computational abilities. It also is increasing due to advice from other National Research Council reports (e.g., NRC 2000, 2002), mandates from the Office of Management and Budget (OMB 2003), and internal EPA guidance (e.g., EPA 1997b). As shown in Box 4-4, there are a number of policy-related questions that can be informed through formal uncertainty analysis.

However, a formal uncertainty analysis, in particular a formal quantitative uncertainty analysis, is difficult to carry out for a variety of reasons. As noted by Mogan and Henrion (1990), "The variety of types and sources of uncertainty, along with the lack of agreed terminology, can generate considerable confusion." In the recent report *Not a Sure Thing: Making Regulatory Choices Under Uncertainty*, Krupnick et al. (2006) noted the lack of a universal typology or taxonomy of uncertainty, making any discussion of the topic of uncertainty analysis for regulatory models difficult. There is also a concern that uncertainty analysis can be difficult to incorporate into policy settings. Krupnick et al. (2006) concluded that one unintended impact of an increased emphasis on uncertainty analysis may be a decrease in decision makers' confidence in the overall analysis. The SAB Regulatory Environmental Modeling Guidance Review Panel (EPA 2006d) elaborates on the concern about using uncertainty analysis in the policy process. Although the panel noted that evaluation of model uncertainty is important in both understanding a system and in presenting results to decision makers, it raised the concern that the use of increasingly complex quantitative uncertainty assessment techniques without an equally sophisticated framework for decision making and communication may only increase management challenges. Further, it is very difficult to perform quantitative uncertainty analyses of complex models, such as regional air quality models (N. Possiel, EPA Office of Air Quality Planning and Standards, personal commun., May 19, 2006). As these complex models are linked to other models, such as those in the state implementation planning process discussed in Chapter 2, the difficulties in performing quantitative uncertainty analysis greatly increases.

Defining Sources of Uncertainty

Although a single uniformly accepted method of categorizing uncertainties does not exist, several general categorizations are clearly defined. As noted by Krupnick et al (2006), the literature distinguishes variability from lack of knowledge and uncertainties in parameters from model uncertainties. Variability represents the inherent heterogeneity that cannot be reduced through additional information, whereas other aspects of parameter uncertainties might be reduced through more monitoring, observations, or additional experiments. The distinction of model uncertainties from parameter uncertainties is also critical. Model uncertainties represent situations in which it is unclear what all the relevant variables are or what the functional relationships among them are. As noted by Morgan (2004), model uncertainty is much more difficult to address than parameter uncertainty. Although identifying and accounting for the consequences of model structural error and uncertainty has only recently become the subject of more sustained and systematic research (Beck 1987, 2005; Beven 2005; Refsgaard et al. 2006), most analyses that have considered the issue report that model uncertainty might have a much larger impact than uncertainties associated with individual model parameters (Linkov and Burmistrov 2003; Koop and Tole 2004; Bredehoeft 2005). Such structural errors amount to conceptual errors in the model, so that if identified at this stage of evaluating the constructed model, assessment should be cast back to reevaluation of the conceptual model.

Krupnick et al. (2006) also identified two other sources of uncertainty important for regulatory modeling: decision uncertainty and linguistic uncertainty. As first observed by Finkel (1990), there are uncertainties that arise whenever there is ambiguity or controversy about how to apply models or model parameters to address questions that arise from social objectives that are not easy to quantify. Issues that fall into this category are the choice of discount rate and parameters that represent decisions about risk tolerance and distributional effects. Uncertainties associated with language, although implicitly qualitative, are important to consider due to the need to ultimately communicate results of a computational model to decision makers, stakeholders, and the interested public. As applied to computational models, sensitivity analysis is typically thought of as the quantification of changes in model results as a result of changes in individual model parameters. It is critical for determining what parameters or processes have the greatest impacts on model results. Figure 4-2 displays the differing interpretations associated with various descriptors that might be used to describe results from models.

Sensitivity and Uncertainty Analysis

Sensitivity and uncertainty analyses are procedures that are frequently carried out during development and application of models. As applied to computational models, sensitivity analysis is typically thought of as the quantification of changes in model results as a result of changes in individual model parameters. The concept of sensitivity analysis has value in the model development phase to establish model goals and examine the advantages and limitations of alternative algorithms. For example, the definition of sensitivity analysis developed by EPA's Council on Regulatory Environmental Models (CREM) includes consideration of model formulation (EPA 2003d). The goal of a sensitivity analysis is to judge input parameters, model algorithms, or model assumptions in terms of their effects on model output. Sensitivity analyses can be local or global. A *local* sensitivity analysis is used to examine the effects of small changes in parameter values at some defined point in the range of these values. A *global* sensitivity analysis quantifies the effects of variation in parameters over their entire space of these values. When addressing global sensitivity, the effect of varying more than one parameter on the response must be considered. A common approach for assessing sensitivity and uncertainty is to run the model multiple times while slightly changing the inputs.

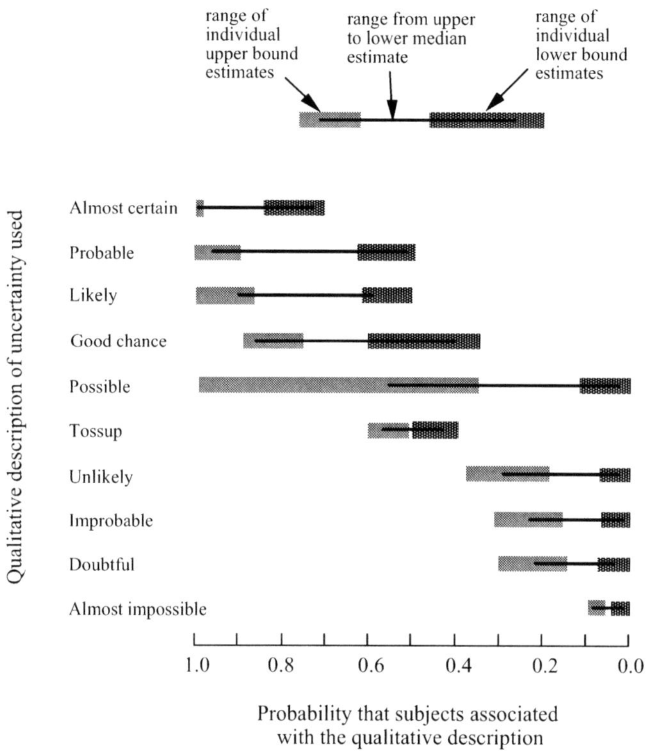

FIGURE 4-2 Range of probabilities that people assign to different words absent any specific context. Source: Adapted from Wallsten et al. 1986. Reprinted with permission; copyright 1986, *Journal of Experimental Psychology: General.*

Quantitative uncertainty analysis is the determination of the variation or imprecision in the output function based on the collective variation of the model inputs using a variety of methods, including Monte Carlo analysis (EPA 1997b). In a broader perspective, uncertainty analysis examines a wide range of quantitative and qualitative factors that might cause a model's output values to vary. All models have inherent capabilities and limitations. The limitations arise because models are simplifications of the real system that they describe, and all assessments using the models are based on imperfect knowledge of input parameters. Confronting the uncertainties in the constructed model requires a model performance evaluation that (1) estimates the degree of uncertainty in the assessment based on the limitations of the model and its inputs, and (2) illustrates the relative value of increasing model complexity, of providing a more explicit representation of uncertainties, or of assembling more data through field studies and experimental analysis.

Model Uncertainty Versus Parameter Uncertainty

Although a distinction between model uncertainty and parameter uncertainty is typically made, there is an argument over whether there is indeed any fundamental distinction. In the sense that both kinds of uncertainty can be handled through probabilistic or scenario analyses, the committee agrees, but notes that this applies only to the uncertainty about the output of models. For assessing uncertainty in model outputs, uncertainty about which model to use can be converted to uncertainty about a parameter value by constructing a new model that is a weighted average of the competing models (e.g., Hammitt 1990). But the issue of selecting a set of models that captures the full space of outcomes and the choice of weighting factors is problematic. Therefore, the committee considers that there is a worthwhile practical distinction between model and parameter uncertainty, if for no other reason than to emphasize that model uncertainty might dwarf parameter uncertainty but can easily be overlooked. This is particularly important in situations where models with alternative conceptual frameworks to the standard model are too expensive to run or do not even exist.

EVALUATION AT THE MODEL APPLICATION STAGE

A new set of practical considerations apply in moving from the development of a computational model to the application of the model to a regulatory problem, including the need for specifying boundary and initial conditions, developing input data for the specific setting, and generally getting the model running correctly. These issues do not detract from the fundamental questions and trade-offs involved in model evaluation. The evaluation will need to consider the degree to which the model is based on generally accepted science and computational methods; whether the model fulfills its designed task; and how well its behavior approximates that observed in the system being modeled. For models that are applied to a specific setting for which the model was developed, these questions should have been addressed at the model development stage, particularly if the developers are the same group applying the model.

However, frequently models are applied by users who are not the developers or even in the same institution as the developers. In many cases, model users might have a choice in the model to use and in alternative modeling approaches. In these cases, model evaluation must address the same fundamental considerations about the appropriateness of the model for the application and explicitly address the trade-offs between the need for the model to get the right answer for the right reason and the need for the modeling process to be transparent to stakeholders and the interested public. The discussion here focuses on the evaluation of model applications using uncertainty analysis. Later in this chapter, we discuss other elements of model evaluation relevant to this stage, including peer review and documentation of the model history. Chapter 5 discusses issues related to model selection.

Uncertainty Analysis at the Model Application Stage

At the model application stage, an uncertainty analysis examines a wide range of quantitative and qualitative factors that might cause a model's output values to vary. Effective strategies for representing and communicating uncertainties are important at this stage. For many regulatory models, credibility is enhanced by acknowledging and characterizing important sources of uncertainty. For many, it is possible to quantify the effects of variability and uncertainty in input parameters on model predictions by using error propagation methods discussed below. They should not be confused with or used in place of a more comprehensive evaluation of uncertainties, including the consideration of model uncertainties and how decision makers might be informed by uncertainty analysis and use the results.

The Role of Probability in Communicating Uncertainty

Realistic assessment of uncertainty in model outputs is central to the proper use of models in decision making. Probability provides a useful framework for summarizing uncertainties and should be used as a matter of course to quantify the uncertainty in model outputs used to support regulatory decisions. A probabilistic uncertainty analysis may entail the basic task of propagating uncertainties in inputs to uncertainties in outputs (which would commonly, although perhaps ambiguously, be called a Monte Carlo analysis). Bayesian analysis, in which one or more sources of information are explicitly used to update prior uncertainties through the use of Bayes' theorem, is another approach for uncertainty analysis and is better, in principle, because it attempts to make use of all available information in a coherent fashion when computing the uncertainties of any model output. However, the committee considers the use of probability to quantify *all* uncertainties to be problematic. The committee disagrees with the notion that might be inferred from such statements as Gayer and Hahn's (2005): "We think policy-makers should design regulations for controlling mercury emissions so that expected benefits exceed expected costs if that statement is interpreted to mean that large-scale analyses of complex environmental and human health effects should be reduced not only to a single probability distribution but also to a single number, the mean of the distribution." Although it is hard to argue with the principle that regulations should do more good than harm, there are substantial problems in reducing the results of a large-scale study with many sources of uncertainty to a single number or even a single probability distribution. We contend that such an approach draws the line between the role of analysts and the role of policy makers in decision making at the wrong place. In particular, it may not be appropriate for analysts to attach probability distributions to critical quantities that are highly uncertain, especially if the uncertainty is itself difficult to assess. Further, the notion that reducing the results of a large-scale modeling analysis to a single number or distribution is at odds with one of the main themes that began this chapter, that models are tools for helping make decisions and are not meant as vehicles for producing decisions. In sounding a cautionary note about the difficulties of both carrying out and communicating the results of probabilistic uncertainty analyses, we are trying to avoid the outcome of having models (and a probabilistic uncertainty analysis is the output of a model) make decisions.

To see the difficulties that can result from this purely probabilistic approach to uncertainty analysis, consider the following EPA study that, in response to an OMB requirement, treated uncertainties probabilistically. In a study on emissions from nonroad diesel engines, one of the key parameters affecting the monetary value of possible regulations was the value assigned to a human life (EPA 2004b). A probability distribution for this parameter was obtained using the following approach. The 5th percentile of the value of a human life was set at $1 million, based on a study that had used this value as the 25th percentile. The 95th percentile was set at $10 million, based on another study that had used this value as the 75th percentile. Then, using "best professional judgment" (see Table 9B-1 in EPA 2004b), a normal distribution was fit using the 5th and 95th percentile points, resulting in the mean value of a human life being $5.5 million. The numbers $1 and $10 million are rough approximations at least in part due to the decimal number system. Nevertheless, despite the arbitrary choice of highly rounded figures for the 5th and 95th percentiles, there is nothing preposterous about $5.5 million as an estimate of the value of a human life (although there is something disconcerting about the fact that this distribution assigns a probability of 0.0083 to the value of a human life being negative). However, the real problem here is not in the details of how this distribution was obtained, but that it was done with the goal of providing policy makers with a single distribution for the net benefit of a new regulation. Though the committee does not imply that such analysis arbitrarily assigns values, monetizing such things as a human life or visibility in the Grand Canyon clearly requires assessing what value some relevant population assigns to them. Thus, it is important to draw the distinction between uncertainties in such valuations and, say, uncertainty in how much lowering NO_x emissions from automobiles will affect ozone levels at some location.

Another approach to uncertainty assessment is to calculate outcomes under a fixed number of plausible scenarios. If nothing in each scenario is treated as uncertain, then the outcomes will be fixed numbers. For example, one might consider scenarios with such names as highly optimistic, optimistic, neutral, pessimistic, or highly pessimistic. This approach makes no formal use of probability theory and can be simpler to present to stakeholders who are not fully versed in probability theory and practice. One advantage of the scenario approach is that many of those involved in modeling activities, including members of stakeholder groups and the public, may attach their own risk preference (such as risk seeking, risk adverse, or risk neutral) to such scenario descriptions. However, even using multiple scenarios ranging from highly optimistic to highly pessimistic will not necessarily ensure that such scenarios will bracket the true value.

In thinking about the use of probability in uncertainty analysis, it is not necessary or even desirable to consider only the extremes of representing all uncertainties by using probability or by not using probability at all. The assessment can have a hybrid approach using conditional distributions in which a small number of key parameters having large, poorly characterized uncertainty are fixed at various plausible levels and then probabilities are used to describe all other sources of uncertainty.

To illustrate how conditional probability distributions can be used to describe the uncertainty in a cost-benefit analysis, consider the following highly idealized problem. Suppose the economic costs of a new regulation are known to be $5 billion with very little uncertainty. Furthermore, suppose that nearly all of the benefit of the regulation will be through lives saved. Thus, to assess the monetized benefits of the regulation, we need to know how many lives will be saved and what value to assign to each life. Suppose that, based on a thorough analysis of the available evidence, the uncertainty about the number of lives saved by the regulation has a median of 1,000 and follows the distribution shown in Figure 4-3a. Furthermore, as in EPA (2004b), assume that the value of a human life follows a distribution with $1 million as its 5th percentile and $10 million as its 95th percentile, but unlike the EPA study, we assume that this distribution follows what is known as a lognormal distribution (rather than a normal distribution), which has the merit of assigning no probability to a human life having a negative value.

This lognormal distribution is shown in Figure 4-3b. If we further make the natural assumption (see footnote) that the number of lives saved and the value of a human life can be treated as statistically independent quantities, then it follows that the distribution of the net benefits of the regulation is given by the distribution in Figure 4-4, which shows that the probability that the net benefit will be positive is slightly under one-fourth, and the expected net cost is approximately $630 million.[1]

---

[1]Cost-benefit analyses are commonly full of (often unexamined) assumptions of statistical independence of various quantities. In the present circumstance, one might try to argue that there is no plausible relationship between the value judgment in monetizing a human life and the uncertainty in the number of lives saved by a regulation, and therefore an assumption of statistical independence is justified. However, depending on the nature of the uncertainty one is attempting to represent through probability, it is possible to envision substantial dependence between the two quantities. In particular, suppose part of the evidence for the number of lives saved is based on laboratory animal studies at high exposure levels and that the number of lives saved thus depends on how one extrapolates from high to low doses and from animals to humans. If the probability distribution of net benefits is supposed to represent the diversity of personal judgments of a set of experts and if experts who tend to make conservative assumptions about how to extrapolate results from animal studies also tend to assign a high value to a human life, the assumption of independence

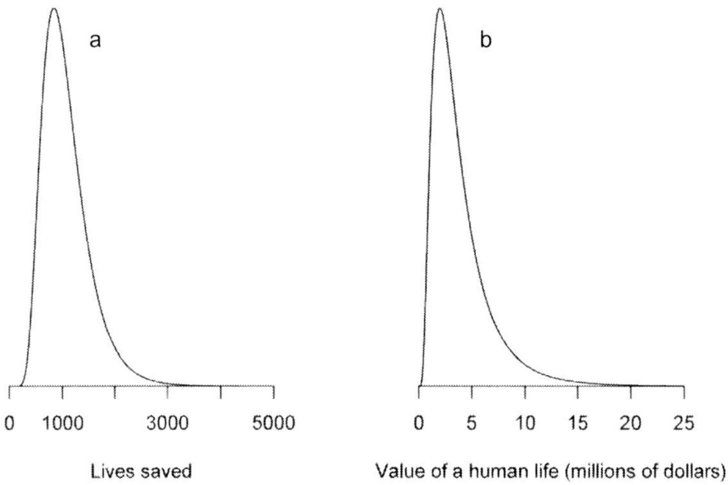

FIGURE 4-3 (a) Hypothetical distribution representing uncertainty in number of lives saved by a policy; (b) distribution representing uncertainty in value of a statistical life.

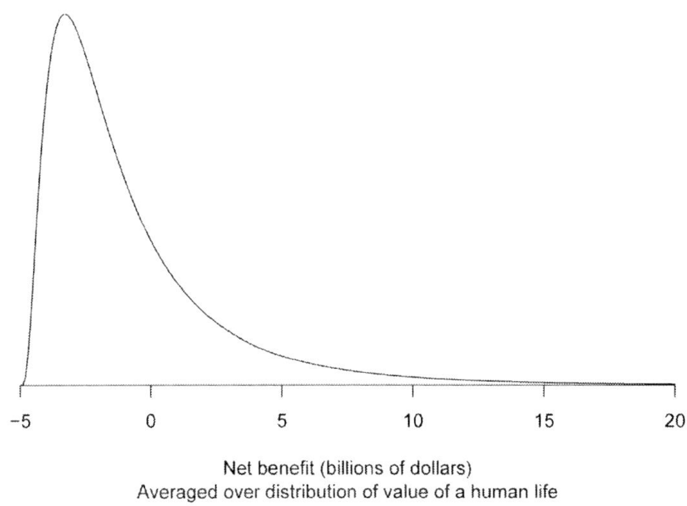

FIGURE 4-4 Unconditional posterior distribution for net benefit of policy.

---

would be violated, and that could have a significant impact on the cost-benefit analysis.

This conclusion is highly sensitive to the difficulty of quantifying the value of a human life. Instead of averaging over the distribution in Figure 4-3(b) for this value, cost-benefit analyses could give the distribution conditional on different values. For example, Figure 4-5 gives the conditional distribution of net benefits when the value of a human life is set at $1 million or $10 million (it also gives the unconditional distribution from Figure 4-4). It is now seen that if the value of a human life is set at $1 million, the probability that the regulation has a positive net benefit is essentially zero, whereas if the value of a human life is set at $10 million, the probability of a positive net benefit is large (about 0.96), the expected net benefit being over $5.8 billion.

We contend that Figure 4-5 is a clear and important improvement of Figure 4-4. Free software providing more sophisticated tools for visualizing conditional distributions is available, for example, in the lattice library for R (Murrell, 2006), or if interactive graphics are desirable, the program XGobi (Swayne et al. 2002) is available. Interactive graphics would allow the policy maker to choose values of one or more key parameters and then view the conditional distribution of the net benefit given these parameters. However, interactive computer programs are no substitute for human interaction, and the committee strongly encourages extensive interaction between scientists and policy makers when policy makers can ask various "what if" questions to help in their decision making.

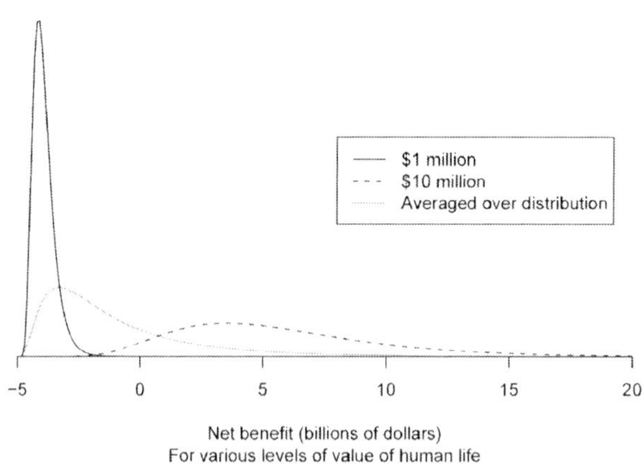

FIGURE 4-5 Conditional distributions of net benefit assuming different amounts for the value of a statistical life.

To use this hybrid approach to uncertainty analysis, the user will have to decide which uncertainties to average over (treat probabilistically) and which to condition on (consider some set of fixed values). Although there cannot and should not be hard and fast rules on this matter, the committee can offer some guidance. As already noted, quantities with large and poorly characterized uncertainties are prime candidates for conditioning. Value judgments, such as the worth of a human life or of high visibility at the Grand Canyon, may often fall into this category. Uncertainties about model choice are another example of an uncertainty that should not be addressed using an expected value. For example, in extrapolating animal studies of toxicity at relatively high doses to much lower doses in humans, conclusions may differ by large factors, depending on the assumptions made. Rather than attempt, via a Bayesian calculation, to average over a host of models that all fit the data about equally well but result in different conclusions about low-dose human toxicity, it may be better to give several possible conclusions under varying assumptions on how to extrapolate across species and doses.

In providing this guidance, it is not the committee's intent to dismiss the considerable amount of work done on monetization of value judgments, nor the work on Bayesian model averaging (Hoeting et al. 1999). The committee is asserting, however, that policy makers should be informed of the impacts of changing assumptions about highly uncertain parameters on an analysis, and the impacts should not be buried in a technical analysis.

In addition to the plots of conditional distributions, other summaries of the uncertainty analysis should be given to decision makers. For example, distributions of quantities other than net benefit, such as those given in Figure 4-3, should be routinely included in the analysis.

Because most probabilistic uncertainty analyses, whether or not explicitly Bayesian, now would calculate distributions of outcomes of interest by using simulations, another approach to conveying the results of an uncertainty analysis would be to provide "typical" sample points from the simulation. For example, terms such as "optimistic" and "neutral" could be defined in terms of percentiles of the outcome distribution. To be specific, suppose "highly pessimistic" means the 5th percentile of the distribution of net benefits, "pessimistic" the 25th percentile, "neutral" the median, "optimistic" the 75th percentile, and "highly optimistic" the 95th percentile. The user could then present a table of key inputs (or intermediate outputs) for the sample points at those percentiles in the simulated distribution of net benefits. Alternatively, summaries of the distributions of key inputs for various ranges of the net-benefit distribution could be used. In effect, the distributions would be conditional of the inputs given the net benefits rather than of the net benefits given certain inputs, as suggested above.

It might be argued that providing multiple summaries that include a combination of conditional distributions, typical sample points, and distributions of intermediate outputs will be too much information for policy makers. However, interviews conducted with former EPA decision makers on the use of uncertainty analysis in regulatory decision making do not support this pessimistic assessment of the quantitative literacy of environmental policy makers (Krupnick et al. 2006). If the uncertainty assessment is clearly presented, with succinct summaries of the major sources of uncertainty and their impacts on the conclusions (including a list of any potential nontrivial sources of uncertainty that were not taken into account), the committee considers that such an uncertainty analysis will help empower decision makers and improve the decision-making process, especially if decision makers are included interactively in the process of putting together the summaries of uncertainty.

Evaluation of Statistical Models

As discussed in various places throughout this report, statistical models often might involve the use of flexible regression models, for instance, using polynomials or splines to characterize the relationship between an exposure and a response of interest. Statistical models are widely used to analyze epidemiological data. Important considerations are adjusting for confounders, handling missing data, accounting for study design, and so forth; therefore, assessing the adequacy of statistical models may involve a complex set of considerations quite distinct from many of the other modeling settings discussed elsewhere in this chapter. Whereas assessing the adequacy of a process-based model, such as a mass-balance model for indoor air pollution, relies on general theory, epidemiological models tend to be specialized and tailored to the specific context at hand and cannot be assessed in the abstract. In contrast to many process-based models, technical aspects of epidemiological models may be simple—for example, relying on simple linear or logistic regression models. For such models, the model development and model application stages are the same.

The challenge, however, is in making sure that all the appropriate information has been incorporated in an adequate manner. For example, has the study design been appropriately reflected in the analysis? If the study population represents a probability-based sample, then it may be important to include sampling weights in the analysis before generalizing results to the full population. Data quality is another important consideration. For instance, if exposure assessment is subject to measurement error, it may be important to adjust for that to avoid bias associated with the error. A critical issue concerns whether the appropriate covariates have been identified so that potential confounding can be tested for and adjusted for. Missing data are an inevitable challenge in even the most well-run epidemiological study, so it is important to assess the impact of missing data, from the loss through the follow-up, to ensure that the analysis is not subject to bias. From a more technical perspective, it is important to ask whether the modeling assumptions are appropriate and whether the chosen model fits the observed data reasonably well. This question might involve assessment of the appropriateness of any linearity assumptions (testing for outliers, for example) and might be assessed by looking at residuals and applying goodness-of-fit tests. Were the appropriate steps taken to identify all the appropriate confounders? Was the method of covariate selection documented? Were covariates incorporated into the model in an appropriate way? If covariates are measured on a continuum, were nonlinearities tested for? Equally important is the question of how the primary exposure of interest was included in the dose-response model. Good model assessment and exploration include considerations of alternatives to the shape of the dose-response curve, exploration of possible lag effects, and so forth. Sensitivity analysis in general is a powerful and highly recommended strategy for ensuring that results are not driven by one or two key assumptions. Finally, it is important to be sure that the statistical software being used is numerically stable and reliable.

There is much literature on residuals analysis and goodness-of fit-analysis for statistical models, and there are a number of popular approaches. One technical concern to be aware of is that caution is needed when assessing model adequacy and goodness of fit using the same data as those used as the basis for fitting the model. There are a number of ways to address this concern. The Akaike information criterion (AIC) and the Bayesian information criterion (BIC) correspond to the estimated log-likelihood plus an additional penalty term that reflects the number of parameters in the model. Both AIC and BIC approaches are popular for assessing goodness of fit and represent the differences between frequentist and Bayesian statistical methods. Formal Bayesian approaches are also possible, of course, in which case examination of posterior and predictive distributions play an important role in model assessment. Ideally, models would be checked against new, independent data. However, this is not always possible.

MANAGEMENT OF THE EVALUATION PROCESS

This section addresses practices for managing the evaluation process. The life cycle of a model can be complex for any single model and immensely difficult when the full range of EPA regulatory models are considered. Thus, the committee offers overarching principles for management of the evaluation process. At its core, the committee sees the need for a strategy for model evaluation (a model evaluation plan) and a description of the model's historical development, use, and evaluation to follow a model throughout its life stages. This recommendation is not intended to be a bureaucratic exercise that relies on extensive documentation. Some model evaluation plans and histories for simple models may be limited. The goal to achieve is a substantive commitment from the agency to ensure that model evaluation continues throughout a model's life. This goal raises the organizational question of accountability and responsibility for such efforts. The committee does not presume to make organizational recommendations, nor does it recommend the level of effort that should be expended on any particular type of evaluation. Because of the great diversity of models, no single approach is likely to be viable. However, EPA needs some mechanism that audits the process to make sure that (1) there is a life-cycle model evaluation plan, (2) there are resources to carry out the evaluation and pay the true costs, (3) the EPA modelers respond to peer reviews, and (3) they follow through in both completing the actions requested in the peer review and in continuing the peer review process. The crucial element is that the process should be a means to an end, namely, a model fit for its purpose and not an end in itself.

The Use of Model Evaluation to Establish Model Acceptability Criteria

The committee discussed the merits of providing a uniform set of scientific and technical acceptability criteria applicable to all regulatory models. It became clear that the range of model types and model applications at a regulatory agency such as EPA will not work under an over-arching set of acceptability criteria except for the requirement that each model be based on methods, science, and assumptions that the agency accepts as appropriate. The committee found that no one had yet established such criteria, although work on this topic by the Netherlands National Institute for Public Health and the Environment (RIVM) has been done and is described at the end of this section. Even if such criteria were available, they might well not be applicable to the many and varied settings of EPA's use of models. In addition, there is an intangible policy context to any choice about the acceptability of a model for a given regulatory setting. Resources, public and stakeholder buy-ins, and other factors can play a role. Regulations are never tied to model capabilities, so there is often an imperfect correlation between model capabilities and regulatory needs.

Acknowledging that this area is a matter for further substantial research, the committee considered what combination of scientific and technical factors and process steps should be considered in developing model acceptability and application criteria. The factors are the following:

- Scientific pedigree
- Model structure and components
- Model capabilities and limitations
- Inputs and outputs
- Applicable space and time scales
- Applicable substances
- Key sensitivities and uncertainties
- Model performance evaluation
- Parsimony
- Peer review

The committee notes that information on these factors should arise from the model evaluation process, it forming the basis for setting acceptability and applicability criteria for specific models and specific model applications. How the above factors are addressed in the model evaluation plan will vary among different model types. The committee envisions that the acceptability and applicability criteria be presented either within the model evaluation plan or in a separate document on the basis of information provided about the factors. We consider below explicit examples of what several of these factors mean and how they relate to acceptability and applicability.

Scientific Pedigree. "Scientific pedigree" is a shorthand term for considering the fundamental science that forms the basis of the conceptual model. The scientific pedigree considers the origin and the quality of the concepts and theories behind the model and each of its constituent parts (Funtowicz and Ravetz 1990). Over the years, the fundamental scientific and other understandings that are used in constructing models have been consolidated and refined to produce—at maturity—a product with a pedigree. The merit in the scientific pedigree concept is that it is applicable, in principle, at various levels, from assessments of an integrated suite of models to its major subblocks (such as atmospheric chemistry and human toxicology) and down to the details of the parameters characterizing the mathematical expression of individual processes.

Model Structure and Components. Those who evaluate the acceptability of a model for a given purpose should see a diagram and brief description of the major components of the model. At one extreme, a model may have multiple models (such as source, transport, exposure, dose, risk, and uncertainty models) linked by managing software. Here, a diagram and a summary of structure and components are essential for judging acceptability and applicability. For example, if an atmospheric transport model is linked to a soil model and a surface-water model, it is important to know how the intermedia transfers from air to soil and air to water are managed, that is, in one or two directions. This information could determine acceptability for specific classes of pollutants. Another example is a one-box pond model that is applicable and acceptable for representing a small surface-water body but might not be applicable and acceptable for representing one of the Great Lakes, where there are potentially distinct subregions within the water body. A third example is regional mass-balance models designed to capture the chemical mass balance for aggregated sources over a large space and time scales but are not designed to capture detailed source-receptor relationships.

Model Capabilities and Limitations. The model evaluation plan should clearly distinguish the capabilities and limitations of a model. For example it will usually be important to identify whether a transport model can handle organic chemicals, inorganic chemicals, or microorganisms; whether an economic impact model is macro or micro in its level of resolution; or whether an exposure model works for the short term (minutes to hours) or the long term (days to years). Another example is when an air dispersion model is used to assess how pollutant emissions translate to concentrations downwind from a source. This type of model is acceptable for modeling the transport of stack emissions but may not be acceptable for modeling such conditions as pesticide drift from field applications or for estimating exposure when the receptor population is indoors or moving in ways that are not captured in the model. Among the issues that should be covered in a statement of capabilities and limitations are the inputs required to run the model, the outputs provided by the model, the space and time scales for which the model applies, the types of substances that the model can address, and a discussion of key sensitivities and uncertainties.

Model Performance Evaluation. It is difficult to imagine that a model is acceptable for a regulatory application without some level of performance evaluation showing that the model matches field observations or at least that its results match the results of another well-established model. Acceptability will to some extent be proportional to the level of performance evaluation. Ideally but rarely, a model will be corroborated using one or more independent sets of field data similarly matched to the model's operating domain. Model-to-model comparisons are useful adjuncts to corroboration and in some cases may be sufficient to establish acceptability in the absence of any relevant field data for model comparison.

Parsimony. In light of its recommendation on parsimony, the committee notes that acceptability and applicability decisions need information about parsimony. For those who must select an appropriate model, it is important to know if and how the model developers addressed the issue of parsimony. In particular, did they start with a high level of detail and reduce detail so long as it had no impact on the model or start with a simple model and add detail to meet performance criteria for "validation" and calibration? What type of sensitivity analysis was used to make this determination? There is also substantial literature in related fields that bears on the issue of how much precision or accuracy is needed to inform regulatory decisions. In law, this literature is referred to as the "optimal precision" literature (e.g., Diver 1983). In economics and risk assessment, the issue is referred to as the "value of information" or VOI approach (Finkel and Evans 1987). In terms of VOI, the choice to make models more detailed depends on the degree to which the more elaborate models are judged likely to improve policy outcomes and on the costs of developing and transitioning to more detailed models. In the committee's view, this choice should also include the impact of any loss of transparency.

Peer Review. In most cases, peer review is essential for acceptability, but the level of peer review depends on the nature of the model and its application. Peer review is also useful for providing details on model applicability. The peer review process can also be used to gather information on other factors discussed here to make a determination of model acceptability and applicability.

As a final point, the model evaluation plan created when the model was developed or the peer review process should provide some statement about when an accepted model is no longer acceptable or in need of updates. Some examples of events that make models no longer acceptable are (1) the model has been shown to produce erroneous results (false positives or false negatives) in important regulatory applications; (2) alternative approaches with higher reliability are available and can be developed without unreasonable costs, including transition costs; and (3) key inputs required by the model are found to be incorrect or out of date—for example, demographic data that are 30 years old and no longer updated.

An example of a systematic approach to scientific and technical acceptability criteria for scientific assessments, including those based on environmental modeling, is shown in the activities of the RIVM Environmental Assessment Agency (RIVM/MNP) in its "Guidance for Uncertainty Assessment and Communication" (van der Sluijs et al. 2003). RIVM/MNP's guidance extends beyond the quantitative assessment of uncertainties in model results to focus on the entire process of environmental assessment. The guidance is composed of a series of interrelated tools, including a mini-checklist and a Quickscan questionnaire that asks analysts in a concise set of questions to reflect explicitly on how the assessment deals with issues related to problem framing, stakeholder participation, selection of indicators, appraisal of the knowledge base, mapping and assessment of relevant uncertainties, and reporting of the uncertainty information (Janssen et al. 2005). Other tools available include a detailed guidance document and a tool catalogue for uncertainty assessment (van der Sluijs et al. 2003, 2004). Underlying the checklist is the philosophy that there is no single metric for assessing model performance, there is no typically "correct" model, and models need to be assessed in relation to particular functions. This philosophy echoes this report's discussion of models as tools. The checklist offers modelers a systematic self-evaluation that should provide some guidance on how the modelers are developing the model. It should also help to determine where and why problems may occur (Risbey et al. 2005).

Developing a Model Evaluation Plan

As discussed earlier in this chapter, model evaluation is a multifaceted activity involving peer review, corroboration of results with data and other information, QA/QC checks, uncertainty and sensitivity analyses, and other activities. Viewed in this way, model evaluation is not a one-time event. Even when a model has been thoroughly evaluated and peer reviewed, new scientific findings may raise new questions about model quality, or new applications may not have been anticipated when the model was originally developed. Further, no two model evaluation plan will be alike. A plan should focus on the bigger picture—that model evaluation is intended to address the appropriateness of a given model for a given application and aid in a model's improvement. This plan and the resources devoted to model evaluation should be commensurate with the scope, detail, and regulatory impacts of the model (for example, the scientific complexity, a new application of an existing model, and the likelihood of an application's influence). This plan might evolve with time and experience, especially for long-lived models.

Such a plan could help address a critical shortcoming with regulatory model evaluation. A random sampling of the models listed in the CREM model database shows that most EPA models provide only limited information on model evaluation, and almost none of the models provide a model evaluation plan. Thus, there is typically no consideration of how long-term model evaluation will occur throughout a model life stages. Under the heading "Model Evaluation" in the CREM database, most models present individual statements, such as

- "Currently undergoing beta-testing and model evaluation…."
- "Code verification, sensitivity analysis, and qualitative and quantitative uncertainty analysis have been performed. The model has been internally and externally peer reviewed."
- "The program and user's manual were internally peer reviewed."
- "The model and user's manual were externally peer reviewed by outside peer reviewers and beta testers. The comments from these testers were reviewed by EPA's Office of Research and Development…."

Some models have been subjected to more extensive model evaluation exercises, and at least one has followed through on a model evaluation plan. To gain some insight on how to develop and carry out a model evaluation plan, we consider two examples of models with implicit and explicit model evaluation plans—CMAQ and TRIM.FaTE. CMAQ, the community multiscale air quality modeling system, which is discussed in previous chapters, has been designed to approach air quality in an integrated fashion by including state-of-the-science capabilities for modeling multiple air quality issues, including tropospheric ozone, fine particles, toxics, acid deposition, and visibility degradation. TRIM.FaTE is a spatially explicit, compartmental mass-balance model that describes the movement and transformation of pollutants over time through a user-defined, bounded system that includes biotic and abiotic compartments (EPA 2003g). The extensive documentation on CMAQ includes discussions on the need for and approaches to model evaluation. For example, at one CMAQ workshop, Gilliland (2003) outlined the elements of the CMAQ model evaluation plan. However, the CMAQ web site and CMAQ documentation does not demonstrate an overall evaluation plan. Although it is clear that a number of model evaluations are performed with CMAQ, they typically seem to be directed toward a single aspect or application of the model. It is difficult to see how the plan's activities were conceived, conducted, and fit into an overall scheme. In contrast to CMAQ (and most other EPA models), the TRIM.FaTE model project includes an explicit model evaluation plan in its initial documentation and in follow-up reports on its website (EPA 2006k). The plan identifies the goals and elements of the model evaluation, including conceptual model evaluation, mechanistic model evaluation, data quality evaluation, structural evaluation, and overall performance evaluation. For each of those elements, the model developers provide details on planned activities and the results of activities that have been carried out. The developers follow up with subsequent model evaluation reports that provide results from each of the elements. For the committee, the TRIM.FaTE model evaluation plan and its execution provides a useful example for how to prepare, conduct, and communicate a model evaluation plan for a model of this complexity and scope. It represents a base-case approach to the type of evaluation plan contemplated in this report. Box 4-5 discusses an additional example of life-cycle evaluation for models assessing the persistence and long-range transport of organic chemicals.

BOX 4-5 Life-Cycle Evaluation of Models for Assessing Persistence and Long-Range Transport Potential

As discussed in the text, the EPA model TRIM.FaTE and the model CMAQ are examples of models that have been subjected to more extensive model evaluation exercises that were initiated early in the model development and continue through to the model dissemination stage. Another example that shows the value in evaluating a model from conceptual through use stages is the work of the Organization for Economic Cooperation and Development (OECD 2004) to develop a screening model for assessing the persistence and long-range transport potential of chemicals. The goal of this effort was a consensus model that was evaluated against a broad set of available models and data. The evaluation process began at a workshop in 2001 where the model performance and evaluation goals were set before model selection and development began (OECD 2002). To act upon the recommendations, an OECD expert group was established in 2002. This group published a guidance document on the use of multimedia models for estimating environmental persistence and long-range transport. From 2003 to 2004, the expert group performed an extensive comparison of nine available multimedia fate and transport models to compare and assess their performance (Fenner et al. 2005; Klasmeier et al. 2006). Following this effort, the expert group developed a parsimonious consensus model representing the minimum set of key model components identified in the model comparison. The expert group then convened three international workshops to disseminate this consensus model and provide an on-going model evaluation forum (Scheringer et al. 2006). In this example, significant effort was invested (more than half of the total effort in the OECD case) in the conceptual and model formulation stages. Moreover, much of this effort focused on performance evaluation. The committee recognizes that each model's life cycle is different but notes that attention should be given to developing consensus-based approaches in the model concept and formulation stages. Conducting concurrent evaluations at these stages in this setting resulted in a high degree of buy-in from the various modeling groups.

The committee recognizes the burden that could be placed on model developers to conceive and audit a model evaluation plan. However, the evaluation plan does not have to be a lengthy report. For simple models, it can be a page or two. The following are key elements of the model evaluation plan:

- An evaluation plan for the life cycle of the model that is commensurate with the nature of the model (for example, scientific complexity, new model or application of an existing model, the likelihood of an application's being influential).
- Describe the model (in general) and explain its intended uses.
- Use a thematic structure or diagram to summarize all the elements of the evaluation plan—in particular, the elements that will be used in different stages of model development and application (elements such as the conceptual model, data, model testing, and model application).
- Discuss the events that could trigger a need for major model revisions or that make the model obsolete. This discussion should be specific to the model in question and could be fairly broad and qualitative, such as discussing new science that makes a current model outdated, new regulations, and substantial errors uncovered. The plan should provide criteria to differentiate the need to make a revision of substance rather than to expend resources unnecessarily on continual minor changes. The list of events triggering the need for a major model revision or that might render a model obsolete should itself be periodically updated.
- Specifically identify responsibilities, accountabilities, and resources (for example, staff time, consultant time, and funding) needed to accomplish elements of the plan.

Model History

Models can be developed and applied over many years. During this time, a large number of people could be involved in various aspects of a model's development, evaluation, and application. Many of these people may contribute to or have a specific interest in relatively few elements of this process. This life history of the model can be lost if experiences with a model are not documented and archived. Without an adequate record, a model may be applied incorrectly, or activities may be undertaken that are repetitive or ignorant of earlier efforts. For example, an expert peer reviewer of a model application needs to understand the full history of the model's evaluation. Has another reviewer evaluated the mathematical algorithms in the original development phase? Has another expert determined that the databases used to develop the model are appropriate? What is the range of environmental parameters for which the model is reasonably accurate and does the new application fall within those parameters? Thus, keeping such a model's history is essential for effective model use. Maintaining a history of significant events regarding the model and a documentation of the model history would support transparency objectives and help modelers use and improve a model long after the original developers are gone and the verbal history is lost. Such a history could include the purpose of the model, major assumptions and modifications, and the history of its use and evaluation.

Peer Review

Peer review is the time-honored way to improve the quality of a scientific product. Experts in the field are the only ones with the capabilities of evaluating highly technical material. Even then, experts may require additional analyses or material to perform a rigorous review. Also, a peer review is only useful if the reviewers' comments are considered and used appropriately to revise the model. The regulatory environment model setting also makes peer review fundamentally different from the review of other scientific products that do not have regulatory applications (Jasanoff 1990, 2004). These complexities are key reasons why a model evaluation plan and why a record of the model's life history are needed.

The tradition of one-time peer review for models is essential but not sufficient. Having knowledgeable peers review the conceptual model could help to identify important issues related to transparency, such as how to explain the model and how to present the results, and whether the scope and impacts considered within the conceptual model are consistent with the regulatory problem at hand. It also could be helpful for models with large regulatory impacts or complex scientific issues to have a periodic peer review or peer advisory process in which the peers interact with the model developers and users throughout the model's life. As noted in EPA's most recent version of its peer review guidance, the agency is beginning to appreciate that obtaining peer review earlier in the development of scientific products might be desirable (EPA 2006a). The agency is also recognizing that multiple peer review events also might be useful, particularly when the work product involves complex tasks, has decision branching points, or is expected to produce controversial findings (EPA 2006a).

Although OMB encourages agencies to have peer reviewers run the models and examine the computer code (Schwab 2004), resources provided to reviewers are usually limited, and individual reviewers typically cannot do extensive testing or code verification. However, adequate peer review of a model may involve reviewers running the model results against known test cases, reviewing the model code, and running the model for an array of problems. It also may demand particular attention to the intended applications of a model, because a model that is well-suited for one purpose at one time may not be appropriate for another purpose at that time or the same purpose at a different time.

A peer review is so basic to model quality and its acceptance that it must be excellent in substance, as well as appearance. Therefore, careful attention must be given to the three foundations of selecting peer reviewers: scientific qualifications, conflicts of interest, and balance of bias. These issues are explained in some detail by EPA (2006a). All reviewers must be, without exception, scientific peers. They should be free of conflicts of interest (for example, the result of the review should not have a direct and predictable impact on the finances of the reviewer), and if that is not possible on rare occasions, they should be publicly justified and explicitly permitted by appropriate agency authorities. After the first two requirements for selecting peer reviewers are met, the peer review committee biases must be balanced. Biases cannot be eliminated because they are based on the experts' perspectives, but a peer review committee should not be biased in any given direction. Finally, a high-quality peer review is the result of EPA's commitment to the overall model evaluation process. More attention should be paid to providing sufficient time and material to the peer reviewers to enable them to fulfill a well-developed charge.

Adequate peer review of a model, especially a very complex model or a model that has a substantial impact on environmental regulations, may involve reviewers running the model results against known test cases, reviewing the model code, and running the model for an array of problems. It is unreasonable to expect such peer reviews to be done without compensation. To obtain such an in-depth peer review, the committee sees the need for support in the form of compensation and perhaps in running the model for conditions that the reviewers specify. The committee considers such peer review to be part of the cost of building and using models, especially models with a large impact on regulatory activities.

Stakeholder Review

Seeking involvement of stakeholders is sometimes seen as merely a legal requirement, which it often is, but a more flexible attitude may take greater advantage of this required process. Fundamentally, stakeholder review helps addresses the social, legal, financial, and political contexts of the designated task. Stakeholders may have information or perspectives that can help guide the process. All of those legitimately holding a stake in the outcome of the process of evaluation will not share the same formulation of the policy problem; nor, given widely differing attitudes toward risk, will they all come to the same conclusion or judgment, even under an identical formulation. The groups involved in the environmental regulatory process can be risk takers, risk averse, and risk managing, to name but three classes of perspective (Thompson 1989). They can be knowledgeable in a classic scientific sense, such as when an affected party has or hires experts, or in a realistic sense, such as when members of the public identify an exposure pathway that was not identified by the experts. These various groups can participate in the model evaluation process through various activities, including producing their own supporting or conflicting model results and challenging the legitimacy or accuracy of a model in public comments or judicial actions. However, to engage stakeholders fully in model evaluation, decision makers must understand the financial, legal, and political risks attached to the outcomes of the regulatory activities (for which the model has been designed); the cultural attitudes of the various stakeholders toward those risks; the ways that stakeholders might use to manipulate the task context; and the extent to which various stakeholders trust the process of model evaluation. Although the committee recognizes that encouraging stakeholder participation adds to the complexity of model evaluation, their involvement may result in a more transparent or more robust model.

Vigorously involving the general public is possible, as demonstrated in agency modeling activities that are site-specific. In designing cleanup plans for Superfund sites, for example, EPA not only must solicit the community's input but also must often convene multiple interviews and educational meetings to provide the community with a sufficient opportunity to respond to agency risk assessments and cleanup proposals (for example, see National Contingency Plan, 40 CFR §

300.430[c])[2] given the local impacts of these model-based regulatory decisions, the general public can invest considerable resources in overseeing the quality of EPA's cleanup models and can even obtain grants to hire technical experts to review EPA's technical assessments (40 CFR, Part 35, Subpart M [Technical Assistance Grants]). Even though the mandatory public participation requirements are relatively similar for diffuse, national issues, the level of involvement by the general public can increase dramatically as the agency's decisions become localized and specific to a particular community.

The special needs of stakeholders should to be considered. Time for review can be a barrier. As mentioned above, stakeholders can have perspectives useful to those involved in the model evaluation process, but they must have time to develop such comments and transmit them to the peer reviewers to be effective. Special attention must be paid to involve stakeholders because most are not technically expert. Some groups may have the scientific staff or the budget to hire consultants to perform model review and often do so from their own perspectives. In contrast, other smaller organizations (for example, small businesses and small environmental advocacy groups) and the general public do not have the resources to comment on regulatory actions that may have a substantial impact on them. Such organizations and individuals must rely on the process to inform them and make recommendations that will protect their interests. However, these processes are typically not at all clear to these individuals and groups.

Thus, buy-in by some stakeholders and the general public may be based on trust of the model evaluation process rather than on the results of the process. Making progress in achieving meaningful peer review of science and models pertaining to regulation may depend more on having stakeholders agree in advance on appropriate methods and evaluation protocols than on subsequent (conventional) scientific peer review. Establishing and demonstrating the reliability and credibility of the peer review process itself is every bit as crucial as the conventional challenge of establishing the reliability and credibility of the information. Dealing effectively with stakeholders and the general public can have collateral benefits. Process transparency may enhance buy-in by stakeholders and the general public, especially if the regulation affects their behaviors, and later by the courts, if challenges are brought against a regulation.

---

[2]This document describes community relations requirements for remedial actions.

Learning from Prior Experiences—Retrospective Analyses of Models

The final issue in managing the model evaluation process is management of the learning that is developed through the examination of prior modeling experiences. Retrospective analysis of models is important for developing improvements to individual models and regulatory policies as well as systematically enhancing the overall modeling field. There have been many examples of retrospective analysis of particular environmental modeling activities. Box 4-6 describes three such examples. However, even with the widespread use of models at EPA, there has been little attempt to generalize prior experiences with models and classes of models into systematic improvements for the future. One reason may be the reluctance by the agency to disclose errors, criticisms, and shortcomings in the adversarial and legally constrained setting that environmental regulatory modeling activities often occur. The discussion of groundwater model retrospective analysis of Bredehoeft (2003, 2005) demonstrates that generalizing prior experiences with models does not necessarily imply the commitment of a great deal of modeling resources but possibly does imply the use of the experiences of veteran modelers to provide insights.

The committee has considered the value of retrospective studies as a critical part of model evaluation from two primary perspectives. The first perspective is broad. It concerns the retrospective evaluation of classes of models—for example, models of groundwater flow, surface water, air pollution, and health risks assessment. The goal of such an approach would be to investigate whether there are systematic weaknesses that are characteristic of various types of models. For example, based on modeling experiences in his past work and work described by other hydrogeologists, Bredehoeft (2003, 2005) estimated that in 20-30% of groundwater modeling efforts, surprising occurrences indicated that the conceptual models underlying the computer models were invalid.

The second perspective is somewhat narrower. If a specific model is being used for several years for high impact issues, its performance for its intended use should be questioned. For such cases, data are probably available for retrospective analyses that were not available at the time of model construction. In addition to data that have been collected over time, other data that are critical to model evaluation may be identified and collected specifically to address the question, "how well does the model work?"

With respect to the question of how well different classes of models work, it would be useful to know whether different classes of models have common weaknesses. As noted, Bredehoeft's work suggests that groundwater models are subject to surprises that show their underlying conceptual models to be invalid. Bredehoeft reported that one suggestion arising from that observation is to carry alternative conceptual models

BOX 4-6 Retrospective Analysis of Model Predictions

Retrospective analysis of environmental regulatory models often occurs when particular model predictions are later compared to measurements or results from other models. Examples include comparisons of estimates of regional light-duty-vehicle emissions and the effectiveness of emission-control policies with those predicted by the MOBILE model, an assessment of an air quality model's ability to simulate the change in pollutant concentrations associated with a known change in emissions, and comparisons of groundwater conditions and containment transport with those predicted by groundwater models.

Light-duty-vehicle emissions inventories are important for a wide range of air quality management activities, including serving as inputs to air quality models as well as direct indicators of the performance of emissions control policies. For regulatory activities outside of California, the MOBILE model is used for regulatory purposes. Methods that have been used for retrospective assessments of MOBILE's vehicle emission estimates include remote sensing of vehicle exhaust emissions, direct emissions measurement at vehicle emissions inspection and maintenance (I/M) stations and other facilities; the use of fuel sales to model emissions; and measured concentrations of air pollutions, both in ambient air or in tunnels, to infer emissions (e.g., Stedman 1989; Fugita et al. 1992; Gertler et al. 1997; Singer and Harley 2000; Watson et al. 2001; NARSTO 2004). A recent NARSTO report on emissions inventories found significant improvements over the past decade in the correspondence of model predictions and observations of on-road emissions inventories, but with significant shortcomings remaining (NARSTO 2005). One particular issue related to MOBILE's estimates of control program effectiveness that has gathered much interest is the comparison of modeled estimates of the benefits of I/M programs in reducing emissions to those estimated through remote sensing and other techniques (Lawson 1993; Stedman et al. 1997; Air Improvement Resources 1999; CARB 2000a; Wenzel 2001). An NRC study of I/M programs concluded that an earlier version of the MOBILE model overestimated emissions benefits (MOBILE5), though the most recent version of the model (MOBILE6) has reduced estimated I/M benefits (NRC 2001a; Holmes and Cicerone 2002).

EPA's Model Evaluation and Applications Research Branch is currently performing a retrospective analysis of the CMAQ model's ability to simulate the change in a pollutant associated with a known change in emissions (A. Gilliland, EPA, personal commun., May 19, 2006, and March 5, 2007). This study, which

EPA terms a "dynamic evaluation" study, focuses on a rule issues by EPA in 1998 that required 22 states and the District of Columbia to submit State Implementation Plans providing NO_x emission reductions to mitigate ozone transport in the eastern United States. This rule, know as the NO_x SIP Call, requires emission reductions from the utility sector and large industrial boilers in the eastern and midwestern United States by 2004. Since theses sources are equipped with continuous emission monitor systems, the NO_x SIP call represents a special opportunity to directly measure the emission changes and incorporate them into model simulations with reasonable confidence. Air quality model simulations were developed for summers 2002 and 2004 using the CMAQ model, and the resulting ozone predictions were compared to observed ozone concentrations. Two series of CMAQ simulations have been developed to test two different chemical mechanisms in CMAQ to consider model uncertainty that is associated with the representation of chemistry in the model. Given that regulatory applications use the model's prediction of the relative change in pollutant concentrations, dynamic evaluations such as these are particularly relevant to the way the model is used.

Groundwater models are critical for regulatory applications, such as assessing containment transport from hazardous waste sites and assessing the long-term performance assessments of high level nuclear waste disposal sites. Bredehoeft (2003, 2005) summarizes a series of post-hoc studies where later observations were used to evaluate how well earlier groundwater modeling did in predicting future conditions. Besides errors in conceptual models of the system, which are discussed in the body of this report, Bredehoeft identified insufficient observations for specifying input parameters and boundary conditions as another critical reason why model predictions did not match observations. An additional issue cited was that, in some instances, the assumed environmental management actions that were modeled ended up to be very different from the actual actions taken. It is important to note that, while the number of studies discussed in Bredehoeft (2003, 2005) was extensive, the modeling resources involved was not. Instead, the insights were developed by having an experienced modeler look across a number of applications for overarching conclusions. This observation is important when considering the resource needs and scope of retrospective analysis.

into an analysis. In his experience, Bredehoeft noted that alternatives are not carried into analysis. However, such an approach has been applied in the health risk assessment area. Distinctly different conceptual models for health risks from sulfur oxides in air were discussed in several papers by Morgan and colleagues (Morgan et al. 1978, 1984). These papers described alternative conceptualizations of the health risks that are incompatible with each other but that, at the time of the analyses, were supported by some data.

In his 2003 paper, Bredehoeft described the following difficulties with conceptual models:

- Modelers tend to regard their conceptual models as immutable.
- Time and again errors in prediction revolve around a poor choice of the conceptual model.
- More often than not, data will fit more than one conceptual model equally well.
- Good calibration of a model does not ensure a correct conceptual model.
- Probabilistic sampling of the parameter sets does not compensate for uncertainties in the appropriate conceptual models or for wrong or incomplete models.

The point of this list is that models with conceptual problems cannot be improved by enhanced efforts at calibration or management of uncertainties. The best chance for identifying and correcting conceptual errors is through an ongoing evaluation of the model against data, especially data taken under novel conditions.

The question that should be explored is whether other classes of models share a common weakness. For example, as a class, what weaknesses would be identified by an evaluation of air dispersion, transport and atmospheric chemistry models, or structure-activity relationships? Identifying systemic weaknesses would focus the attention on the most productive priorities for improvement. With a long-term perspective, there will be cases in which it is possible to compare model results with data that were not available when the models were built.

A key benefit of retrospective evaluations of models of individual models and of model classes is the identification of priorities for improving models. Efforts to add processes and features of diminishing importance to current models may be of much lower benefit than revisions based on priorities derived from retrospective analyses. The committee did not identify a solid technical basis for deciding whether specific models should be revised other than to address the perception that a specific model was incomplete.

RECOMMENDATIONS

The committee offers several recommendations based on the discussion in this chapter. They deal with life-cycle model evaluation, peer review, uncertainty analysis, retrospective analysis, and managing the model evaluation process.

Life-Cycle Model Evaluation

Models begin their life cycle with the identification of a need and the development of a conceptual approach, and proceed through building of a computational model and subsequent applications. Models also can evolve through multiple versions that reflect new scientific findings, acquisition of data, and improved algorithms. Model evaluation is the process of deciding whether and when a model is suitable for its intended purpose. This process is not a strict verification procedure but is one that builds confidence in model applications and increases the understanding of model strengths and limitations. Model evaluation is a multifaceted activity involving peer review, corroboration of results with data and other information, quality assurance and quality control checks, uncertainty and sensitivity analyses, and other activities. Even when a model has been thoroughly evaluated, new scientific findings may raise unanticipated questions, or new applications may not be scientifically consistent with the model's intended purpose.

Recommendations

Evaluation of a regulatory model should continue throughout the life of a model. In particular, model evaluation should not stop with the evaluation activities that often occur before the public release of a model but should continue throughout regulatory applications and revisions to the model. For all models used in the regulatory process, the agency should begin by developing a life-cycle model evaluation plan commensurate with the regulatory application of the model (for example, the scientific complexity, the precedent-setting potential of the modeling approach or application, the extent to which previous evaluations are still applicable, and the projected impacts of the associated regulatory decision). Some plans may be brief, whereas other plans would be extensive. At a minimum each plan should

- Describe the model and its intended uses.
- Describe the relationship of the model to data, including the data for both inputs and corroboration.
- Describe how such data and other sources of information will be used to assess the ability of the model to meet its intended task.

- Describe all the elements of the evaluation plan by using an outline or diagram showing how the elements relate to the model's life cycle.
- Describe the factors or events that might trigger the need for major model revisions or the circumstances that might prompt users to seek an alternative model. These could be fairly broad and qualitative.
- Identify responsibilities, accountabilities, and resources needed to ensure implementation of the evaluation plan.

It is essential that the agency is committed to the concept that model evaluation continues throughout a model's life. Model evaluation should not be an end unto itself but a means to an end, namely, a model fitted to its purpose. EPA should develop a mechanism that audits the evaluation process to ensure that an evaluation plan is developed, resources are committed to carry it out, and modelers respond to what is learned. Although the committee does not make organizational recommendations or recommendations on the level of effort that should be expended on any particular type of evaluation, it recognizes that the resource implications for implementing life-cycle model evaluation are potentially substantial. However, given the importance of modeling activities in the regulatory process, such investments are critical to enable environmental regulatory modeling to meet challenges now and in the future.

Peer Review

Peer review is an important tool for improving the quality of scientific products and is basic to all stages of model evaluation. One-time reviews, of the kind used for research articles published in the literature, are insufficient for many of the models used in the environmental regulatory process. More time, effort, and variety of expertise are required to conduct and respond to peer review at different stages of the life cycle, especially for complex models.

Recommendations

Peer review should be considered, but not necessarily performed, at each stage in a model's life cycle. Some simple, uncontroversial models might not require any peer review, whereas others might merit peer review at several stages. Appropriate peer review requires an effort commensurate with the complexity and significance of the model application. When a model peer review is undertaken, EPA should allow sufficient time, resources, and structure to assure an adequate review. Reviewers should receive not only copies of the model and its documentation but also documentation of its origin and history. Peer review for some regulatory models should involve comparing the model results with known test cases, reviewing the model code and documentation, and running the model for several types of problems for which the model might be used. Reviewing model documentation and results is not sufficient peer review for many regulatory models.

Because many stakeholders and others interested in the regulatory process do not have the capability or resources for a scientific peer review, they need to be able to have confidence in the evaluation process. This need requires a transparent peer review process and continued adherence to criteria provided in EPA's guidance on peer review. Documentation of all peer reviews, as well as evidence of the agency's consideration of comments in developing revisions, should be part of the model origin and history.

Quantifying and Communicating Uncertainty

There are two critical but distinct issues in uncertainty analysis for regulatory environmental modeling: what kinds of analyses should be done to quantify uncertainty, and how these uncertainties should be communicated to policy makers.

Quantifying Uncertainty

A wide range of possibilities is available for performing model uncertainty analysis. At one extreme, all model uncertainties could be represented probabilistically, and the probability distribution of any model outcome of interest could be calculated. However, in assessing environmental regulatory issues, these analyses generally would be quite complicated to carry out convincingly, especially when some of the uncertainties in critical parameters have broad ranges or when the parameter uncertainties are difficult to quantify. Thus, although probabilistic uncertainty analysis is an important tool, requiring EPA to do complete probabilistic regulatory analyses on a routine basis would probably result in superficial treatments of many sources of uncertainty. The practical problems of performing a complete probabilistic analysis stem from models that have large numbers of parameters whose uncertainties must be estimated in a cursory fashion. Such problems are compounded when models are linked into a highly complex system, for example, when emissions and meteorological model results are used as inputs into an air quality model.

At the other extreme, scenario assessment and/or sensitivity analysis could be used. Neither one in its simplest form makes explicit use of probability. For example, a scenario assessment might consider model results for a relatively small number of plausible cases (for example, "pessimistic," "neutral," and "optimistic" scenarios). Such a deterministic approach is easy to implement and understand. However, scenario assessment does not typically include information corresponding to conditions not included in the assessment and whatever is known about each scenario's likelihood.

It is not necessary to choose between purely probabilistic approaches and deterministic approaches. Hybrid analyses combining aspects of probabilistic and deterministic approaches might provide the best solution for quantifying uncertainties, given the finite resources available for any analysis. For example, a sensitivity analysis might be used to determine which model parameters are most likely to have the largest impacts on the conclusions, and then a probabilistic analysis could be used to quantify bounds on the conclusions due to uncertainties in those parameters. In another example, probabilistic methods might be chosen to quantify uncertainties in environmental characteristics and expected human health impacts, and several plausible scenarios might be used to describe the monetization of the health benefits. Questions about which of several plausible models to use can sometimes be the dominant source of uncertainty and, in principle, can be handled probabilistically. However, a scenario assessment approach is particularly appropriate for showing how different models yield differing results.

Communicating Uncertainties

Effective decision making will require providing policy makers with more than a single probability distribution for a model result (and certainly more than just a single number, such as the expected net benefit, with no indication of uncertainty). Such summaries obscure the sensitivities of the outcome to individual sources of uncertainty, thus undermining the ability of policy makers to make informed decisions and constraining the efforts of stakeholders to understand the basis for the decisions.

Recommendations

Quantifying Uncertainty

In some cases, presenting results from a small number of model scenarios will provide an adequate uncertainty analysis (for example, cases in which the stakes are low, modeling resources are limited, or insufficient information is available). In many instances, however, probabilistic methods will be necessary to characterize properly at least some uncertainties and to communicate clearly the overall uncertainties. Although a full Bayesian analysis that incorporates all sources of information is desirable in principle, in practice, it will be necessary to make strategic choices about which sources of uncertainty justify such treatment and which sources are better handled through less formal means, such as consideration of how model outputs change as an input varies through a range of plausible values. In some applications, the main sources of uncertainty will be among models rather than within models, and it will often be critical to address these sources of uncertainty.

Communicating Uncertainty

Probabilistic uncertainty analysis should not be viewed as a means to turn uncertain model outputs into policy recommendations that can be made with certitude. Whether or not a complete probabilistic uncertainty analysis has been done, the committee recommends that various approaches be used to communicate the results of the analysis. These include hybrid approaches in which some unknown quantities are treated probabilistically and others are explored in scenario-assessment mode by decision makers through a range of plausible values. Effective uncertainty communication requires a high level of interaction with the relevant decision makers to ensure that they have the necessary information about the nature and sources of uncertainty and their consequences. Thus, performing uncertainty analysis for environmental regulatory activities requires extensive discussion between analysts and decision makers.

Retrospective Analysis of Models

EPA has been involved in the development and application of computational models for environmental regulatory purposes for as long as the agency has been in existence. Its reliance on models has only increased over time. However, attempts to learn from prior experiences with models and to apply these lessons have been insufficient.

Recommendations

The committee recommends that EPA conduct and document the results of retrospective reviews of regulatory models not only on single models but also at the scale of model classes, such as models of groundwater flow and models of health risks. The goal of such retrospective evaluations should be the identification of priorities for improving regulatory models. One objective of this analysis would be to investigate systematic strengths and weaknesses that are characteristic of various types of models. A second important objective would be to study the processes (for example, approaches to model development and evaluation) that led to successful models and model applications.

In carrying out a retrospective analysis, it might be helpful to use models or categories of models that are old by current modeling standards, because the older models could present the best opportunities to assess actual model performance quantitatively by using subsequent advances in modeling and in new observations.

Models and Rule-makings

The sometimes contentious setting in which regulatory models are used may impede EPA's ability to implement some of the recommendations in this report, including the life-cycle evaluation process. Even high-quality models are filled with components that are incomplete and must be updated as new knowledge arises. Yet, those attributes may provide stakeholders with opportunities to mount formal challenges against models that produce outputs that they find undesirable. Requirements such as those in the Information Quality Act may increase the susceptibility of models to challenges because outside parties may file a correction request for information disseminated by agencies.

When a model that informs a regulatory decision has undergone the multilayered review and comment processes, the model tends to remain in place for some time. This inertia is not always ideal: the cumbersome regulatory procedures and the finality of the rules that survive them may be at odds with the dynamic nature of modeling and the goal of improving models in response to experience and scientific advances.

In such an adversarial environment, EPA might perceive that a rigorous life-cycle model evaluation is ill-advised from a legal standpoint. Engaging in this type of rigorous review may expose the model to a greater risk of challenges, at least insofar as the agency's review is made public, because the agency is documenting features of its models that need to be improved. Moreover, revising a model can trigger lengthy administrative notice and comment processes. However, an improved model is less likely to generate erroneous results that could lead to additional challenges, and it better serves the public interest.

Recommendations

It is important that EPA institute best practice standards for the evaluation of regulatory models. Best evaluation practices may be much easier for EPA to implement if its resulting rigorous life-cycle evaluation process is perceived as satisfying regulatory requirements, such as those of the Information Quality Act. However, for an evaluation process to meet the spirit and intent of the Information Quality Act, EPA's evaluation process must include a mechanism for any person to submit information or corrections to a model. Rather than requiring a response within 60 days, as the Information Quality Act does, the evaluation process would involve consideration of that information and response at the appropriate time in the model evaluation process.

To further encourage evaluation of models that support federal rulemakings, alternative means of soliciting public comment on model revisions need to be devised over the life cycle of the model. For example,

EPA could promulgate a separate rule-making that establishes an agency-wide process for the evaluation and adjustment of models used in its rules. Such a programmatic process would allow the agency to provide adequate opportunities for meaningful public comment at important stages of the evaluation and revision of an individual model, without triggering the need for a separate rule-making for each revision. Finally, more rigorous and formalized evaluation processes for models may result in greater deference to agency models by interested parties and by reviewing courts. Such a response could decrease the extent of model challenges through adversarial processes.

Model Origin and History

Models are developed and applied over many years by participants who enter and exit the process over time. The model origin and history can be lost when individual experiences with a model are not documented and archived. Without an adequate record, a model might be incorrectly applied, or developers might be unable to adapt the model for a new application. Poor historical documentation could also frustrate stakeholders who are interested in understanding a model. Finally, without adequate documentation, EPA might be limited in its ability to justify decisions that were critical to model design, development, or model selection.

Recommendations

As part of the evaluation plan, a documented history of important events regarding the models should be maintained, especially after public release. Each documentation should have its origin with such key elements as the identity of the model developer and institution, the decisions on critical model design and development, and the records of software version releases. The model documentation also should have elements in "plain English" to communicate with nontechnical evaluators. An understandable description of the model itself, justifications, limitations, and key peer reviews are especially important for building trust.

The committee recognizes that information relevant to model origins and histories is already being collected by CREM and stored in its model database, which is available on the CREM web site. CREM's database includes over 100 models, although updating of this site has declined in recent years. It provides information on obtaining and running the models and on the models' conceptual bases, scientific details, and results of evaluation studies. One possible way to implement the recommendation for developing and maintaining the model history may be to expand CREM's efforts in this direction. The EPA Science Advisory Board review of CREM contains additional recommendations with regard to specific improvements in CREM's database.

5

Model Selection and Use

The last and perhaps most important stage of the life cycle of a regulatory model is its application to an environmental regulatory issue. How a model arrives at the point of application and how much of its development is specific for a given application vary greatly. For example, modelers who develop a model for a specific application may also apply it, while others who develop a general model do not use it for a particular application. Box 5-1 describes one such model.

The objective of this chapter is to describe issues that arise in selecting models for their applications in regulatory activities. As done throughout this report, regulatory activities considered include any case for which EPA uses a model to aid in developing regulations, such as setting standards, or for which EPA or others develop plans to implement or enforce regulatory requirements. The ultimate goal for all applications is to use all available and appropriate information when selecting a model. In some cases, a model that gets updated on a regular (but not a frequent) schedule might be more appropriate to use, if the updates incorporate information important to the outcome, than to change for the sake of change. However, some degree of stability and predictability is of value to regulators and affected parties. In all cases, evaluation of the model-selection decision assesses the appropriateness of a model or group of models for a specific application. As described in the previous chapter, this assessment involves addressing whether the model is based on generally accepted science and computational methods, whether approximates the behavior observed in the system being modeled.

BOX 5-1 Example of a Generic Model for Application to Specific Settings

A description of one example of model application information is found at EPA's Support Center for Regulatory Atmospheric Modeling web site (EPA 2006*l*). One modeling system described on the site, AERMOD, was developed by the American Meteorological Society and the EPA Regulatory Model Improvement Committee. The AERMOD system is a steady-state plume model that simulates dispersion of air pollutants from point or area sources. It is a good example of an extensively documented model targeted at a broad range of users for regulatory purposes. The AERMOD modeling system includes extensive documentation, including model code, a user's guide, supporting documents, and evaluating databases, all of which are available on the web site of the EPA Support Center for Regulatory Atmospheric Modeling. The supporting documents include details of the model formulation and evaluation, comparison of regulatory design concentrations, an implementation guide (information on recommended use of the model for particular applications), evaluation of model bias, sensitivity analysis, a parameterizations document and peer review document. The evaluation databases include input and output data for model evaluation. User's manuals include instructions for novice and experienced users, decision makers, and programmers. The model code and supporting documents are not static but evolve to accommodate the best available science.

ISSUES IN MODEL SELECTION AND APPLICATION

Model developers and regulators must evaluate how appropriate an existing model is for a specific setting and whether the assumptions and input data are relevant under the conditions of the application. Optimally, a model is applied to a problem within the model-specific application domain near the time of model development. However, frequently, this is not feasible. Thus, models need to be evaluated in context with each application, the degree of evaluation being commensurate with the case. A number of issues arise when selecting and applying a model or a set of models for environmental regulatory activities. These issues are discussed below and include the following: the selection of a model from multiple possibilities, the level of expertise, the assumptions and range of applicability, the cost and availability, the adaptability of the model; and the data availability.

Model Selection

The committee recognizes the wide variability in the availability of alternative modeling approaches for specific regulatory applications.

Thus, guidance on model selection varies. For example, EPA recognizes a single model, the MOBILE model, for developing motor-vehicle emissions inventories for state implementation plans and other air quality regulatory activities outside of California. Although EPA provides guidance for implementing this model, including a user's guide (EPA 2003f) and policy guidance (EPA 2004f), no guidance is needed to select the model. For air quality models, several models, each with its own strengths and weaknesses, might be selected for regulatory activities (Russell and Dennis 2000). The community of air quality modelers is highly specialized and relatively small, and the selection of models is often based on familiarity. In contrast, there are many models from which to select for air dispersion modeling. EPA has developed a guidance document, called Appendix W, on selection of models and on models approved for use (70 Fed. Reg. 68218 [2005]). The guidance is described in more detail in Box 5-2. The EPA Center for Subsurface Modeling Support supports the identification and selection of appropriate subsurface models and supports the review of site-specific modeling efforts at Superfund sites and other large hazardous waste contamination sites (Burden 2004). As with air dispersion modeling, there are many models from which to select; the Center for Subsurface Modeling Support distributes public domain software for over 25 models. There is also a wide range of models possible for performing total maximum daily load (TMDL) analysis (Shoemaker 2004; Wool 2004). Furthermore, fundamentally different modeling approaches are called for, depending on whether the TMDL focuses on the runoff of a pollutant from the watershed, on where a nonpoint source nutrient loading model would be needed, or on whether the TMDL focuses on the concentration of a pollutant in a body of water where a water quality model would be needed.

For all cases that have multiple models available, users must consider many factors when deciding on the most appropriate model to use. These factors include complexity of the problem setting, types of pollutants, spatial and temporal scales, data availability, costs of controls, and an array of practical considerations (for example, available expertise and familiarity) Although no single method for developing a model selection tool would be applicable for the range of conditions faced by regulatory modelers, the recently completed Science Advisory Board's review of the EPA Council on Regulatory Environmental Models (CREM) recommends that the CREM database present competing models in a comparative matrix in the form of a side-by-side comparison table, such as seen in the vehicle sales industry (EPA 2006d).

BOX 5-2 Appendix W: EPA's Guidelines on Air Quality Models

The guidelines, first published in April 1978, was developed to ensure consistency and standardization of model applications for air quality regulations. The guidelines was written in an effort to balance consistency and accuracy in selecting appropriate models. This document, available via the web (70 Fed. Reg. 68218 [2005]; http://www.epa.gov/scram001/guidance/guide/appw_05.pdf), is intended for use by all parties (for example, EPA, state, and local agencies and industry) for calculating the concentration of criteria air pollutants. The guidelines attempts to provide some guidance to model selection while maintaining enough flexibility to account for the complexity and individuality of sources. This document is continuously developed to include new models and updated information on existing or older models and to respond to public comments.

Recommendations concern preferred models, databases, requirements for concentration estimates, use of measured data instead of model estimates, and model evaluation procedures. In some cases, specific models are prescribed for a particular application; in other cases, a type of model is specified. Deviation from the guidelines must be fully supported and documented.

Model selection issues can be further illustrated by considering the use of statistical models for assessing dose-response relationships. The case of EPA's selection of a model for arsenic in drinking water, which is discussed in Chapter 1, provides a good example. In that case, when empirical statistical models in a suite were applied to the data, they differed substantially in their fitted values, especially in the critical low-dose area, which is so important for establishing the benchmark dose used to set a reference dose (see Figure 1-3). This problem highlights the dilemma of model selection in the face of different models with different results. One solution is the use of Bayesian model averaging (BMA) as a tool that avoids having to pick one particular model by combining a class of suitable models. This option, discussed in Box 5-3, is preferable to forcing the choice of a model that may have the best "fit" but that may sacrifice parsimony or that may not account for uncertainty in this case. However, Finkel (2004) described problems with model averaging, and the use of such an approach must be considered on a case-specific basis.

Another approach is to use multiple models of varying complexities to simulate the same phenomena. Using multiple models in such a manner might allow insights into how sensitive results are to different modeling choices and how much trust to put in results from any one model. Box 5-4 shows an example of this approach.

BOX 5-3 Arsenic in Drinking Water: Model Selection

Morales et al. (2000) analyzed the Taiwanese data using a suite of relatively simple empirical models that differed according to how age and exposure were incorporated. All the models assumed that the number of cancers observed in a specific age group of a particular village followed a Poisson model with parameters, depending on the age and village exposure level. Linear, log, polynomial, and spline models for age and exposure were considered. These various models differed substantially in their fitted values, especially in the critical low-dose area; which is so important for establishing the benchmark dose (BMD) used to set a reference dose (RfD). The fitted-dose response model was also strongly affected by whether Taiwanese population data were included as a baseline comparison group. The estimates of the BMD and associated lower limit (BMDL) varied by over an order of magnitude, depending on the particular modeling assumptions used.

This highlights a major challenge for regulatory purposes, namely, which model to base decisions on. One strategy would be to pick the "best" model—for example, use one of the popular statistical goodness of fit, such as the Akieke information criterion (AIC) or the Bayesian information criterion (BIC). These approaches correspond to picking the model that maximizes log-likelihood, subject to a penalty function reflecting the number of model parameters, thus effectively forcing a trade-off between improving model fit by adding addition model parameters versus having a parsimonious description. In the case of the arsenic risk assessment, however, the noisiness of the data meant that many of the models explored by Morales et al. (2000) were relatively similar in terms of statistical goodness-of-fit criteria. In a follow-up paper, Morales et al. (2006) argued that it was important to address and account for the model uncertainty, because ignoring it will underestimate the true variability of the estimated model fit and, in turn, overestimate confidence in the resulting BMD and lead to "risky decisions" (Volinsky et al. 1997). Morales et al. suggest the use of Bayesian model averaging (BMA) as a tool that avoids the need to pick one particular model by combining over a class of suitable models. In practice, estimates based on a BMA approach tend to approximate a weighted average of estimates based on individual models, the weights reflecting how well each individual model fits the observed data. More precisely, these weights can be interpreted as the probability that a particular model is the true model, given the observed data. The figures below show the results of applying a BMA procedure to the arsenic data. Figure 5-1a plots individual fitted models, the width of each plotted line reflecting the weights. Figure 5-1b shows the estimated overall dose-response curve (solid line) fitted via BMA. The shaded area shows the upper and lower limits (2.5% and 97.5% tiles) based on the BMA procedure. The dotted lines show upper and lower limits based on the best fitting models. Figure 5-1b (L30) effectively illustrates the inadequacy of standard statistical confidence intervals in characterizing uncertainty in settings where there is substantial model uncertainty. The BMA limits coincide closely with the individual curves at the upper level of the dose-response curve where all the individual models tend to give similar results.

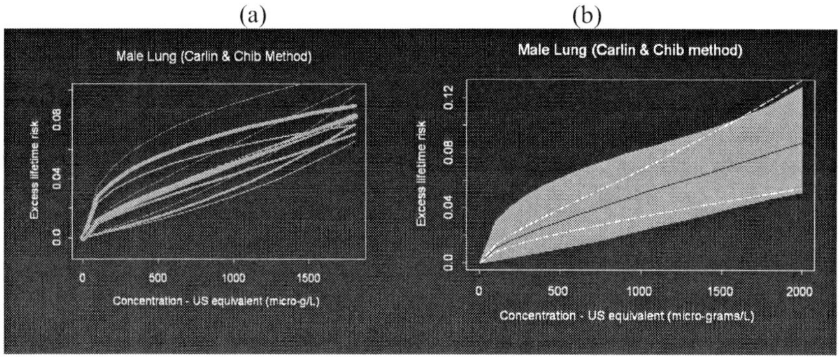

FIGURE 5-1 (a) Individual dose-response models, and (b) overall dose-response model fitted using the Bayesian model averaging approach. Source: Morales et al. 2000.

Model Expertise

Problems can arise if a model is applied incorrectly by an inexperienced user who does not understand how the model operates or who uses the model outside its range of applicability. Such model use would result in potentially erroneous conclusions. Models for regulatory applications will inevitably be used by individuals and groups who are not modelers and who might not be sufficiently trained to catch subtle or even obvious errors. This observation emphasizes the need for training. Box 5-5 mentions two of many ways EPA attempts to improve modeling expertise inside and outside the agency.

Model Documentation and Transparency

For an appropriate model to be selected, both by those assessing whether it would be appropriate for a given case and by those reviewing that decision, it must have adequate documentation for both potential users and those that might scrutinized the model selection decision. Documentation needs, including those related to accepted uses and model origin and history, have been discussed in Chapters 3 and 4. Documenting models for examination by stakeholders and the public provides transparency to build confidence in modeling results (see Box 5-6).

BOX 5-4 Use of Multiple Models of Varying Complexity for Estimating Mercury in Fish

A potential benefit of the clean-air mercury rule, which requires reductions in mercury emissions from coal-fired power plants, is the reduction of human exposure and related health impacts from methylmercury by reducing concentrations of this toxin in fish. There are many challenges and uncertainties in understanding the impact of reductions in atmospheric mercury emissions on human health. In its assessment of the benefits and costs of this rule, EPA used multiple models to look at one particular issue—how changes in atmospheric deposition would affect mercury concentrations in fish—and applied the models to assess some of the uncertainties in this impact (EPA 2005e).

EPA based its national-scale benefits assessment on results from the mercury maps (MMaps) model. This model assumes that there is a linear, steady-state relationship between atmospheric deposition of mercury and mercury concentrations in fish and thus assumes that a 50% reduction in mercury deposition rates results in a 50% decrease in fish mercury concentrations. In addition, MMaps assumes instantaneous adjustment of aquatic systems and their ecosystems to changes in deposition. Thus, there is no time lag in the conversion of mercury to methylmercury and its bioaccumulation in fish. MMaps also does not deal with sources of mercury other than those from atmospheric deposition. Despite those limitations, the agency concluded that no other available model was capable of performing a national-scale assessment.

To further investigate fish mercury concentrations and assess the effects of MMaps assumptions, EPA applied more detailed models, including the spreadsheet-based ecological risk assessment for the fate of mercury (SERAFM) model, to five well-characterized ecosystems. As opposed to the steady-state MMaps model, SERAFM is a dynamic model that calculates the temporal response of mercury concentrations in fish tissues to changes in mercury loading. It includes multiple land-use types for representing watershed loadings of mercury through soil erosion and runoff. SERAFM partitions mercury among multiple compartments and phases, including aqueous phase, abiotic participles (for example, silts), and biotic particles (for example, phytoplankton). Comparisons of SERAFM's predictions with observed fish mercury concentrations for a single fish species in four ecosystems showed that the model underpredicted mean concentrations for one water body, overpredicted mean concentrations for a second water body, and accurately predicted mean concentrations for the other two. The error bars for the observed fish mercury concentrations in these four ecosystems were large, making it difficult to assess the accuracy of the models. Modeling of the four ecosystems also showed how assumed physical and chemical characteristics of the specific ecosystem affected absolute fish mercury concentrations and the length of time before fish mercury concentrations reached steady state.

Although EPA concluded that the best available science supports the assumption of a linear relationship between atmospheric deposition and fish mercury concentrations for broad-scale use, the more detailed ecosystem modeling demonstrated that individual ecosystems were highly sensitive to uncertainties in model parameters. The agency also noted that there were many model uncertainties that could not be quantified. Finally, although the case studies cover the bulk of the key environmental characteristics, extrapolating the individual ecosystem case studies to account for the variability in ecosystems across the country indicated that they might not represent extreme conditions that could affect how atmospheric deposition of mercury would affect fish mercury concentrations in a water body.

This example illustrates the usefulness of investigating a variety of models at varying levels of complexity. A hierarchical modeling approach, such as that used in the mercury analysis, can provide justification for simplified model assumptions or can potentially provide evidence for a consistent bias that would negate the assumption that a simple model is appropriate for broad-scale application.

Resource Requirements and Availability

Model selection must consider whether a model is economically feasible or readily available to potential users. Very complex, detailed models may be expensive to develop and execute. A National Research Council report (NRC 2001c) on the TMDL program urged modelers and decision makers to recognize that simpler analysis can support informed decision making and that complex modeling studies should be pursued only if necessary based on the complexity of the problem. This report recognized that the cost of maintaining and updating a complex model should be considered in model selection, as these costs become cumbersome over time. A possible solution noted by the NRC report is to develop simpler models with existing data that can be iteratively expanded as more data become available.

BOX 5-5 Model Training and Support

EPA has created support networks for aiding in the application of some environmental regulatory models. One of the networks is the Center for Subsurface Modeling Support located within EPA's National Risk Management Research Laboratory. This center provides public domain groundwater and vadose zone modeling software and guidance documents to a variety of users, including universities, state and federal governments, and the private sector (Burden 2004). It also provides training and education. For regional air quality modeling, EPA has created the Community Modeling and Analysis System (CMAS) Center at the University of North Carolina at Chapel Hill. This center is intended to help to promote the use and understanding of the Models-3 air quality modeling suite, including through training courses. The courses are open to everybody (including federal and state employees and scientists from the private sector and academia), although they do assume some prior modeling and computing proficiency. The CMAS center also offers online tutorials.

BOX 5-6 Confidence Building in Models Through Transparency

Placing data sources, software, and the exact list of commands used to produce the model output, along with a good amount of documentation, on a public web site can help build confidence in the specific use of a model. iHAPSS, the internet-based Health and Air Pollution Surveillance System, developed at the Johns Hopkins University, provides an example of this kind of resource.

Even if stakeholders choose not to replicate analyses, which will generally be the case, the presence of such documentation of model use will help to convince stakeholders that the analysts are not trying to hide anything. In some circumstances, some input data may be proprietary or involve privacy concerns or the data set may be too large, thus making this approach unworkable; otherwise, such a public web site should be the norm in high-stakes settings.

A possible concern about making data and code so readily available is that it will make it easier for stakeholders to slow up the decision-making process by raising narrow objections to the large number of choices that are inevitably made in using complex models. Although the committee would still support a high level of openness even if that concern were valid, it is not clear that it is valid. For example, making all data available in readily usable form makes it much easier for others to do their own analyses. As a consequence, criticisms of the form "You should have tried this" or "You need to account for the effect of that" become less cogent because the availability of the data allows the reply, "If you think that matters so much, why not do the analysis yourself?" Making the code available further lowers the barrier to others' modifying an analysis. Thus, greater availability of data and code may help discussions about the appropriateness of a model application to focus on the issues that do matter as opposed to laundry lists of issues that might be conceived to matter.

Assumptions and Specified Range of Use

Understanding the major assumptions and the range of applicability of a model is critical for selection because the assumptions and applicability define an application niche for a model. For example, atmospheric dispersion models typically assume steady, horizontally homogeneous wind fields instantaneously over a given spatial area and are usually limited to 50 km from the source. The use of such a model would not be appropriate for an application at hundreds of kilometers from the source. For the nonsteady-state dispersion model CALPUFF, which allows the model documentation to include information on modeling domains, meteorological data, terrain and land use data, sources, receptors, and modeling options used to develop the model (EPA 2006m). To further demonstrate its application niche to potential users, this model documentation includes a comparison of the modeling results to observations for long-range transport field experiments. All models come with such assumptions and application limitations. Although modelers often have no choice but to use a model for an application in which a major assumption within the model is directly violated or is close to being violated, such an application must be made clear to those who might use or review model results.

Another difficult issue is whether to use a model developed for a different purpose with different specifications. As discussed in Chapter 3, it is often desirable from the standpoint of time and resource investment to use or modify an existing model for a new setting than to develop a new model. However, at what point do the differences make the model inadequate? Professional judgment is required in such cases, and such judgments should differentiate clearly between scientific considerations and other considerations. From an evaluation standpoint, it is critical to make such a decision transparent so it can be commented on and potentially challenged.

Data Availability and Interpretation

A final issue relevant for model selection is the availability and interpretation of data. As discussed in Chapter 3, the mismatch between data needs and availability can result in failure of the model exercise even when the model itself may be a good fit for an individual application. Issues concerning data that come up at model application that are not faced during model development include the need to set boundary and initial conditions, develop site-specific input data, and have access to local monitoring data to test model estimates against observations. Data collection can also aid in reducing uncertainty, improving existing models, and informing developers on when a monitoring program might be useful in reducing uncertainty and simplifying the model. However, models typically can use more data than are available for developing input or for corroborating results with observations. The lack of data requires the use of parameter defaults that are not based on site-specific data.

One approach to lessen the concerns over relying on default parameters in regulatory modeling is to use a tiered approach in which conservative defaults are initially used, possibly with conservative screening models. If a potential problem is detected with conservative defaults, analysis with site-specific data might then be used, possibly with a more refined model, for more refined analysis. Data collection takes time, which may conflict with regulatory time lines, and resources, which may conflict with other priorities. Such conflicts should be explicitly dealt with rather than used as a broad excuse not to collect data. As discussed in Chapter 3, adaptive approaches with iterations among model development and applications and with data-collection efforts are key to improving overall model quality.

Model Extrapolation

Model use in the environmental regulatory process may involve applying a model to extrapolate from conditions that have corroborating information to conditions that have little or no corroborating information available. For example, it might be necessary to extrapolate laboratory animal data to assessments of possible human effects or to extrapolate the recent history of global environmental conditions to future conditions. In these circumstances, uncertainties about the form of a model and of the parameters in any specific model may yield large uncertainties in model outputs.

In some cases, it is clear when application of a model involves extrapolation beyond the data or assumptions used to construct or fit the model. For example, one of the major sources of controversy in the EPA's arsenic risk assessment was the use of a model based on Taiwanese data to estimate risk for the U.S. population (see Box 2-1 in Chapter 2). In this case, model results from one population are extrapolated to another population with differences in genetics, diet, health status, and other factors that could affect the risk relationship (NRC 1999a). In such cases, it is helpful to be as transparent as possible with respect to implicit assumptions that might have an impact on the appropriateness of the extrapolation. In the case of arsenic, for example, extrapolating the Taiwanese results to the U.S. setting involved decisions on whether to use a multiplicative or an additive risk model, as well as assumptions on the typical amount of daily water consumption by individuals in the two countries. Making such assumptions explicit opens the way to sensitivity and uncertainty analyses that can provide a realistic assessment of the impact of applying models to settings outside the context within which they were developed. Extrapolating far beyond available data used to develop the model also puts a particular premium on ensuring that the model's theoretical basis, the processes included in the model, and the selected parameter values within the model are as sound as possible.

Another example in which the use of a model involves extrapolation beyond data or assumptions occurs when EPA forecasts the results of policy decisions into the future. Under such circumstances, EPA often applies models to forecast the impact of regulations over time horizons of years to decades, sometimes incorporating demographic forecasts and forecasts of economic activities (usually from other agencies) as well as assuming that other conditions, such as regulatory and legislative mandates, do not change in the future. Once again, careful sensitivity analyses are needed to assess the impact of various implicit and explicit modeling assumptions to provide a realistic assessment of the uncertainty associated with extrapolating results into the future. This type of approach has been quite effectively applied in the climate change arena where graphs are shown that predict possible future scenarios under a variety of different modeling assumptions. In this sense, the problem of extrapolation beyond the setting in which a model has been developed can be mapped into the broader issue of assessing model adequacy and sensitivity.

If a quantitative structure activity relationship (QSAR) model, which is used to predict physical-chemical properties and environmental fate and transport properties from the chemical structure of a new compound, is being used, the term inside or outside the "domain" is used to indicate whether a model is extrapolated beyond conditions for which the model was constructed or calibrated. The concept of a domain of applicability was one of the six principles developed at a conference of modelers in Setubal, Portugal for use in determining whether a QSAR model is acceptable for chemical management, such as for priority setting, risk assessment, and classification and labeling (Jaworska et al. 2003).

In many applications, extrapolating "far" from known data and conditions is clearly being done. For example, when models are used to predict along a continuum of time, space, or dose, it is clear when the model has moved beyond a point where information is available. In other applications, the model produces output that is not easily placed along a continuum, so it is not clear how much of an extrapolation is being performed. For example, if the model output is the total cost of a regulation and the data are numbers of deaths and pollution levels in cities across the country as well as the per person value of life, the model output can be thought to depend on many unverifiable assumptions. It is in some sense an extrapolation, but it is hard to measure how "far" the output is from the data. Again, this problem puts a premium on ensuring that the model and input parameters are developed on a sound theoretical basis and that the impacts of important assumptions can be assessed through sensitivity and uncertainty analysis.

Specifying Uncertainty

At the model application stage, it is important to have effective strategies for representing and communicating uncertainties. For many regulatory models, credibility is enhanced by acknowledging and characterizing important sources of uncertainty and by acknowledging how uncertainty limits the value of a model as a "truth generator." Modelers should take care to estimate, quantify, and communicate uncertainties accurately to users and regulators. Any limitations in temporal or spatial scales should be stated clearly. The quality of the input data and the resulting limitations on the range of use for the model should be explained in terms of the intended use of the model. Sensitivity to alternative inputs or assumptions should be documented.

As discussed in Chapter 4, interactive graphics would allow the policy maker to choose values of one or more key parameters and then view the conditional distribution of the net benefit, given these parameters. However, interactive computer programs are no substitute for human interaction, and the committee strongly encourages extensive interaction between scientists and policy makers when policy makers can ask various "what-if"-type questions to help them think through their decisions. Policy makers need to be informed explicitly of the impacts of changing assumptions about highly uncertain parameters in a technical analysis; these impacts should not be buried in the analysis.

Communication of Models to Decision Makers

As discussed earlier in this report, models can be best viewed as tools providing input into decisions rather than as truth-generating machines that make decisions. The implications of this finding are clear. Although policy makers may desire an answer from a model, a bright line per se, models are best considered to be one source of input into the regulatory process. The challenge then is to communicate model results and improve the education of policy makers about the capabilities and limitations of the models.

The focus of this effort is typically the EPA policy makers, but it can also include stakeholders who use a model to provide information to EPA, stakeholders who want independent evaluation of the utility of a regulatory model, or even members of the public who must decide whether to change behaviors or to take other actions based on model results. Most of these individuals have one thing in common—they are not technically expert modelers. However, most expert modelers are not expert in issues surrounding decision making. How then can this gap be bridged? One method is to continue to improve model accessibility. Accessibility motivates the committee's recommendation regarding the maintenance of a model's history, including a "plain English" guide to the model. It also motivates the committee's recommendation to continue to improve the transparency of modeling for regulatory decision making, including through web-based tools.

Decision makers should be involved in each stage of model development and use. Their involvement in all aspects of model use, from problem formulation and development of the model's conceptual basis to its application, is fundamental to the appropriate use of models. Such involvement requires successful communication between modelers and decision makers, with emphasis on "between" rather than on "from" one to the other. Both parties have responsibilities to teach and to learn. For major decisions, these responsibilities often must be carried out under tight time constraints in a controversial atmosphere. The modelers need to do more than describe the processes used (for example, peer and stakeholder reviews). They need to describe the modeling elements in an understandable way to a nonexpert. For such communications, it is more about the elements of the model than the precise algorithm used. As described by Voltaggio (2004) when discussing the role of an EPA deputy regional administrator in understanding modeling analysis, the typical questions asked by such decision-makers are related to the assumptions in the model, the quality of the inputs, and the sensitivities of the model results to uncertainties in inputs and other factors. In such cases, decision makers may be relatively ignorant of the model's inner workings.

It also is important for modelers to involve decision makers in the development of uncertainty analysis to ensure that decision makers incorporate their policy expertise and preferences into such assessments. Visualization techniques can be very useful to communicate with decision makers and others, especially when probabilistic approaches are used. However, as noted by Morgenstern (2005), a large body of research on decision makers shows that the manner in which uncertainty information is presented can affect its interpretation. One conclusion that came from interviews with EPA decision makers was the need for more contextual information to accompany any graphic or tabular representations of model uncertainties (Krupnik et al. 2006).

Communication between modelers and a single decision maker can be valuable for all who participate in the regulatory process. The translation of a model from highly technical to more common usage language for an EPA official, for example, can be used by all interested parties. An accessible model evaluation plan helps all.

PROPRIETARY MODELS

At the point of model selection, a regulatory agency may decide to use a proprietary model. A model is proprietary if any component that is a fundamental part of the model's structure or functionality is not available for free to the general public. Components include source code, mathematical equations, input data, user interfaces, or supplemental third-party software (excluding operating systems or development software). Components may also include assumptions or computational methods. A model under copyright is not necessarily proprietary if the model is freely available in its entirety.

The argument for using proprietary models is that, without meaningful intellectual property protections, modelers in the private sector would not have incentives to develop sophisticated models. The arguments against using proprietary models in the regulatory arena have been articulated by environmental groups and industry groups (Sass 2004; Slaughter 2004). Proprietary models to these stakeholders are directly at odds with the goals of open government and transparency.

Motivations for Keeping Information Proprietary

In some cases, proprietary models are used because one might happen to provide the most reliable and dependable output for a specific application. Efforts should be taken to use an open-source model when available; however, model developers might be motivated to maintain the proprietary nature of the models that they develop. These motivations include profit from selling, updating and maintaining the model, training users on the model, and protecting trade secrets.

The best way for a modeler to protect his intellectual investment in a model is to claim trade-secret protection. Protection is immediate and is accomplished by insisting that the model and its contents are secret. There are two main difficulties in evaluating the legitimacy of a trade-secret claim on proprietary models in terms of whether it ultimately serves the public interest. First, the owner of the proprietary information has the best information concerning whether there is a legitimate competitive advantage to keeping the information secret, thus, making it hard for outsiders to evaluate, especially if the owner has other, overlapping reasons to insist on confidentiality (such as to avoid controversy over assumptions and to retain control over the running of the model). Second, it is difficult to evaluate empirically whether providing secrecy to model developers will spur innovation. In other words, would modelers still develop models for the marketplace with private dollars, even without trade-secret protections?

Proprietary Aspects of Environmental Models

The CREM guidance defines a proprietary model as one in which the source code is not universally shared (EPA 2003d). However, a model can also be classified as proprietary if any component of the model is proprietary, including the source code, the input data, or third-party software. These three components are explained in Box 5-7.

The committee heard presentations on three case histories of proprietary models. The Integrated Planning Model (IPM) is a long-term capacity expansion and production costing model for analyzing the electric power sector. It was developed by ICF International and is used by EPA and a wide variety of other groups to assess environmental regulatory activities that affect this sector (Napolitano and Lieberman 2004). The model is used because of its detailed representation of the system, including rich representations of dispatch decisions, capacity expansion, and emission-control options. A key element of the proprietary nature of this model is the thorough representation of the electricity sector. The DEEM-CALANDEX models are used widely to estimate multiple-pathway human exposure models for pesticides. These models were developed by Exponent Inc. The key proprietary feature of these models is their user-friendly interface and ability to do multiple

BOX 5-7 Proprietary Components of Environmental Models

1. *The source code:* This code defines the fundamental function of the model, for example, the computational solution of the mathematical equations representing the underlying theory or code that defines the auxiliary features of the model (for example, graphic user interface). Generally, an executable file will be provided to the user when the source code is not available. Because most if not all environmental models have an underlying theory that is well known, the first type of code will probably only occur if the model owner considers the computational solution (including any assumptions) to be of value, for example, if it is complex or novel and would require considerable effort to duplicate. The second type of code is a more common scenario with environmental models because "usability" is an important value-added feature when the theory and computational solutions are widely known (for example, Gaussian plume models and Calendex).
2. *Input information:* Input information might be kept proprietary if it is confidential and/or if the information management is believed to be value added. An example of the latter case would be when the information is public but the volume of information is large, is frequently updated, and/or requires extensive processing or conversion prior to model use. This is one important proprietary aspect of the Integrated Planning Model.
3. *Third-party software:* All models to some extent rely on third-party proprietary software because they run on an operating system, and most require a specific language compiler and interpreter, both of which typically require a license. All of these are developed for diverse purposes and not specifically for regulatory applications. In other words, there is usually proprietary software used in model *development* (for example, historical languages, such as FORTRAN, but also contemporary applications with "development environments," such as Excel, ArcView, and Analytica). Somewhat in contrast are third-party programs that are necessary for model *use*, for example, to facilitate model analysis or perform common mathematical operations. Examples of third-party software for model use are

- Numerical solvers (for example, to solve systems of ordinary differential equations or linear programming problems).
- Statistical analysis packages (for example, Excel and SAS).
- Database software (for example, Access and Oracle).
- Visualization and analysis software (for example, ArcView).

The line is gray between third-party software for development versus use, since a model can be developed in a specific application with the end use of that application in mind (for example, a GIS model).

analyses quickly (Petersen 2004). The TRANSIMS model predicts vehicle travel on highways and then that information is used as input into mobile-source emissions models. Currently, TRANSIMS is a research model developed by Los Alamos National Laboratory for the U.S.

Department of Transportation. IBM has been hired to commercialize this model by developing user interfaces that will allow users to develop model input, run the model, and visualize the output (Ducca 2004).

Alternatives for Using Proprietary Models

It might be risky to ignore the purported benefits of proprietary models if they appear to be playing an important role in advancing the science of modeling. However, the committee notes that distrust of proprietary models was shared by both a representative of an environmental organization (Sass 2004) and a representative of a pro-business group advocating for regulatory reform (Slaughter 2004). Agencies such as EPA could use a range of alternatives to justify the use of proprietary models, to provide some oversight of these models' reliability, and to limit the potential use of such models. The objective of these alternatives is to have the rigor applied to EPA-developed models also be applied to external models used by EPA in the regulatory process. Many of these issues were discussed by Napolitano and Lieberman (2004) and Petersen (2004).

- An agency could bargain for the added right to disclose publicly the contents of the model. In this case, the agency would pay the model owner to give up his right to keep the information secret. The problem is that the model owner may charge a very large fee to transfer ownership of the model.
- An agency could require the model owner to justify the claim that the model must be kept proprietary. The model owner would be expected to explain the competitive losses from divulging the model, and this justification would be publicly available.
- If the model is ultimately kept confidential, the agency could require the owner to agree to a limited number of "confidentiality agreements" with an objective peer review panel (to be named later) that would evaluate the model and provide a public report on its findings without disclosing the trade-secret-protected information. This process ensures rigorous peer review without releasing protected information.
- Before using a proprietary model, an agency should justify why it is superior to the alternatives. Under such a policy, proprietary models would be disfavored and used only when the agency can provide a compelling justification for doing so.

- An agency could insist on an expiration date for the secrecy protections on the model.
- An agency could insist that the modeler obtain a patent rather than protect its property interest through trade-secret protections (a patent requires the public dissemination of the contents of the model). The problem with obtaining a patent is that it can take years to obtain, and it is also uncertain that some models can be patented. Apparently, copyright protections will work as long as the model is embedded in software, but they will not apply to the underlying ideas in a model (like the algorithms), and thus copyrighting is not a viable means of protecting intellectual property in models.

RECOMMENDATIONS

Proprietary Models

A model is proprietary if any component that is a fundamental part of the model's structure or functionality is not available for free to the general public. The use of proprietary models in the regulatory process can produce distrust among regulated parties and other interested individuals and groups because their use might prevent those affected by a regulatory decision from having access to a model that may have affected the decision. There are many ways in which a model can be proprietary, and some are more prone to engender distrust than others. For example, a model that uses proprietary algorithms may cause more concern than a model that uses publicly available algorithms but has a proprietary user interface.

Recommendations

The committee recommends that EPA adopt a preference for nonproprietary software for environmental modeling. When developing a model, EPA should establish and pursue a goal of not using proprietary elements. It should only adopt proprietary models when a clear and well-documented case has been made that the advantages of using such models outweigh the costs in lower credibility and transparency that accompanies reliance on proprietary models. Furthermore, proprietary models should be subject to rigorous quality requirements and to peer review that is equivalent to peer review for public models. If necessary, nondisclosure agreements could be used for experts to perform a thorough review of the proprietary portions of the model. The review process and results could then be made public without compromising proprietary features. General-purpose proprietary software (for example, Excel, SAS, and MATLAB) usually will not require such scrutiny, although EPA should be cognizant of the costs that obtaining and using such software may impose on interested parties.

Extrapolation

Model use in the environmental regulatory process may involve using the model to extrapolate beyond conditions for which the model was constructed or calibrated or conditions for which the model outputs cannot be verified. For example, it might be necessary to extrapolate laboratory animal data to assessments of possible human effects or to extrapolate the recent history of global environmental conditions to future conditions. In these circumstances, uncertainties about the form of a model and the parameters in the model might yield large uncertainties in model outputs. This problem can be compounded by making a model more complex if the additional processes in the more complex model are unimportant; any extra parameters that need to be estimated could degrade the confidence in the estimates of all parameters.

Recommendations

Extrapolating far beyond the available data for the model draws particular attention in the evaluation process to the theoretical basis of the model, the processes represented in the model, and the parameter values. When critical model parameters are estimated largely on the basis of matching model output to historical data, care must be taken to provide uncertainty estimates for the extrapolations, especially for models with many uncertain parameters.

6

Future Modeling Issues

Modeling will continue to have a central role in future environmental regulatory activities. This is because models are at the nexus of science and policy (Gilman 2005). Critical for this endeavor will be how models incorporate the ever-increasing amounts of observations of natural and human processes and environmental impacts. Vast new measurement programs in fields as diverse as genomics to earth observation systems at scales from the nano to the global pose significant opportunities and challenges for modeling. Although observations alone can influence policy, it is the analysis of this information with models that will allow the full realization of the importance of these measurement programs.

Environmental regulatory modeling also will be greatly influenced by new scientific understandings and enhanced modeling technologies. The potential to incorporate greater understanding of environmental processes, such as the creation of airborne particulate matter from gaseous precursors and the physiological and pharmacokinetic absorption, disposition, metabolism, and excretion of a chemical in the body, is already offering great improvements to modeling capabilities. However, the new information and capabilities come at a time of increasing demand for greater scrutiny of regulatory activities by stakeholders and the public. Thus, improving environmental regulatory modeling does not necessarily imply using the most complex models. New modeling technologies, including developing modular modeling codes or user-friendly programming languages, also can improve modeling transparency and can better match complexity needs to computational tools.

EXPANSION OF MEASUREMENT SYSTEMS

The relationship of models to measurements has been a critical issue throughout the history of modeling. The rapid increase of information about environmental processes, human-environment interactions, and human and environmental impacts brings new challenges to this relationship in the future. The spectrum of new information that will be available to the environmental regulatory process is vast and beyond the scope of this report. Two examples are discussed to indicate the diverse sources of information that have the potential to be available to modeling.

One end of the spectrum could be considered the genomics revolution, which has enabled the analysis of all the genes in a cell at the DNA, mRNA, protein, or metabolite level (NRC 2006b). These tools can be used to better understand the susceptibility of individuals or subpopulations to chemicals, as well as their responses to chemicals (toxicogenomics). For example, genomics tools provide a means to examine changes in gene expression and to examine how these indicators might be used to understand human health impacts (EPA 2004g). Although the capability to understand the potential for toxicants to impact human genes has been present for many years, the innovation of high throughput testing technologies has profoundly expanded the capability to better measure genomic changes (NRC 2006b). The dramatically increasing amounts of information from genomic technologies have spawned a new science called infomatics to enable orderly analysis of vast data sets. Infomatics includes a wide variety of statistical and other computational models at the "research" level rather than at the "regulatory" level at this time. However, substantially more sophisticated computational toxicology methods, including the use of computational models of biological systems and phenomena, will be needed to link genomics data to quantitative estimates of human health risks before the full potential for this information will be realized (NRC 2006b).

Another end of the spectrum of measurement systems that will influence regulatory modeling is the rapid increase in data from environmental satellites and weather data (Foley 2005). The information from these systems provides a truly global climate observation system as well as highly resolved spatial and temporal observations of meteorological phenomenon (Bates 2004). Such measurements may help to discern information on climatic variability, water resources, ecosystem changes, air pollution episodes, and a wide array of other possible applications. Although the sheer volume of data creates unprecedented challenges for data-handling operations, a more fundamental challenge is the scientific use of this information (Kahn 1997).

IMPROVEMENTS IN MODEL METHODS AND TECHNOLOGIES

As with the wide range of new measurement systems that are potentially available, a wide range of modeling approaches and technologies are increasingly applied in the environmental regulatory setting. Again, the spectrum of possible technologies and methods is vast and beyond the scope of this report. The committee discusses two areas as examples: integrated environmental modeling approaches and user-friendly modeling technologies.

One area is the increasing development of integrated modeling approaches. A major difference between "today's" approach and "tomorrow's" approach may be that high-quality models can enable an assessor to describe computationally with reasonable accuracy the relationships depicted in Figure 2-1 in Chapter 2—from source emissions and human activities that give rise to these emissions to adverse outcomes. The continuum from sources to human health responses in the human health risk assessment paradigm is described in many sources (e.g., NRC 1983, 1994; Lioy 1990) and demonstrated in the approach taken by the National Research Council committee in developing research priorities for airborne particulate matter (NRC 1998, 1999b, 2001d, 2004c). Recent advances in modeling tools have greatly enhanced the capabilities to perform computationally intensive multiscale source-to-dose and exposure assessment for a wide range of environmental contaminants (Foley et al. 2003). For example, Georgopoulos et al. (2005a,b) described an integrated source-to-dose modeling framework for assessing population exposures to fine particulate matter, ozone, and air toxics that links emissions, meteorological, air quality, exposure, and dosimetry models. The use of integrated modeling approaches for the environment is not confined to the human health risk assessment field.

Other examples of such integrated environmental modeling approaches that are emerging can be found in the following fields:

- Watershed modeling—The BASINS modeling framework includes watershed nutrient loading and transport models and instream water quality models that operate with a geographical information system (EPA 2001d).
- Risk assessment—The TRIM.FaTE model is a multimedia compartmental model to help assess multimedia chemical fate, transport, and exposure and risk of pollutants in the ambient environment (Efroymson and Murphy 2001; EPA 2003g).
- Hazardous waste risk assessment—The multimedia, multipathway, and multireceptor exposure and risk assessment (3MRA) model can assess potential human and ecological health risks using transport, fate, exposure, and toxicity (EPA 2003h).
- Global change fields—These models link models of energy-economic processes to environmental models (e.g., Rotmans 1990; Holmes and Ellis 1999) and models that link air quality, weather, and climate (Jacobson 2001; Liao et al. 2003, 2004).

These integrated modeling frameworks are typically written in a modular form, as discussed in Chapter 3, which allows users to easily add or remove parts of the model to tailor individual applications to the problem at hand. Software platforms, such as the framework for risk analysis in multimedia environmental systems (FRAMES), are often used to link models and databases under one integrated system. Typically, a user interface facilitates such development.

However, the ever-larger and more-sophisticated models may not necessarily make better regulatory tools. Clarke (2004) and Perciasepe (2005) raise the possibility that pursuing larger and more-sophisticated models make them less and less able to be evaluated and more impenetrable to the public and decision makers.

Other modeling technologies have attempted to improve transparency and build a stronger bridge to the public and decision makers through the use of user-friendly graphic simulation software. One approach is to utilize object-oriented programming languages that allow individual components of a model to be visually and mathematically linked in a user environment that displays how different elements of a model interrelated and that allows users to easily modify the relationship among components. One use of this approach has been in conflict resolution over water resources. Known as share-vision modeling, it involves the common development of a single model or modeling framework by a diverse group of stakeholders involved in a water resources issue facilitated by object-oriented programming software (Lund and Palmer 1997). This approach has been recommended by the Institute for Water Resources as a way to bridge the gap between the specialized water models and the human decision process (Werick 1995).

CHANGES IN PERSPECTIVES ON MODEL USE IN REGULATORY DECISION MAKING

The use of models in the regulatory process in the future also may be affected by changing perspectives of decision makers on the most effective way to use them. Two general approaches are weight-of-evidence and adaptive management strategies. The weight-of-evidence approach has been used long before the original National Research Council's "Red Book" on risk assessment practices (NRC 1983), although definitions and methods for carrying out weight-of-evidence analyses vary (Weed 2005). However, all definitions in the modeling setting recognize that models cannot be used to define a precise "bright line," for example, between attainment and nonattainment of ambient environmental standards. Dolwick (2005) described how the regional air quality modeling community evolved from using models to define in an absolute sense whether a location's emissions reduction plans will result in attainment of National Ambient Air Quality Standards (NAAQS) to using models in a weight-of-evidence approach as the primary element in a suite of tools that includes emissions and air quality monitoring. EPA (2006n) described the agency's guidance on implementing the weight-of-evidence approach for ozone, fine particulate matter, and regional haze standards. The Air Quality Management (AQM) Work Group, which is composed of stakeholders from state and local governments and some industry and nonprofit organizations, endorsed the weight-of-evidence approach as a way to reduce reliance on modeling data as the centerpiece for air quality attainment demonstrations and increase the use of monitoring data and analyses of monitoring data (AQM Work Group 2005). Although the weight-of-evidence approach appropriately recognizes that models are not "truth generators," it must be used in an unbiased manner so that, for example, it is no more likely to be used to relax regulatory requirements than to strengthen them, even when modeling uncertainties cut both ways (NRC 2004a).

Adaptive strategies recognize the importance of improving environmental management strategies as new measurements and modeling analyses become available. Although providing a single definition for such terms as "adaptive management" and "adaptive implementation" suffers from the same problem as defining weight of evidence, some environmental regulatory activities clearly recognize an adaptive approach in which management strategies are later modified based on new modeling, measurements, and research. For example, the Clean Air Act calls for the NAAQS for each criteria pollutant to be reviewed periodically to consider recent scientific findings. The objective of this review is to decide whether the current NAAQS for that pollutant should be revised. Although the process of reviewing and implementing changes in the standards is cumbersome and has not been kept up with the 5-year review cycle mandated in the legislation, the history of the Clean Air Act has seen important revisions to air quality standards as a result of these reviews. Another example is in the cleanup of large mining megasites, where the amount and wide distribution of contaminated materials preclude complete remediation with traditional cleanup approaches envisioned under the Superfund Act. EPA recognizes that many contaminated mining megasites will require operation and maintenance in perpetuity (EPA 2004h). Under conditions where remediation is a long-term process involving many separate projects, some of which cannot be specified at the outset, the agency is forced into an adaptive approach requiring periodic progress reviews and adjustments to unsuccessful remedies. An NRC report focusing on mine-related contamination in the Coeur D'Alene River Basin mining megasite recommended that EPA establish a rigorous, adaptive management process for such mining megasites, a process having well-defined performance milestones, monitoring strategies, and evaluation criteria (NRC 2005a; Gustavson et al. 2007). A final example of an adaptive strategy in environmental regulatory activities is the California Air Resources Board (CARB) process for periodic review and revision, if necessary, of California motor-vehicle emissions standards (NRC 2005c). Because of the far-reaching and long-term nature of the California standards, CARB committed to a biennial review of its motor-vehicle emissions standards program to monitor manufacturer compliance plans, to identify any problems with the feasibility of its demanding program, and to modify the standards if deemed necessary (e.g., CARB 1994, 2000b). This process resulted in modifications to California's standards, most notably to its zero emissions vehicle mandate (CARB 2004).

IN CLOSING

Models have a prominent future in the environmental decision-making process because their value clearly outweighs their inherent imperfections. The use of environmental regulatory models in the future will have to deal effectively with the vastly increasing amounts of data, improvements in modeling methods and technologies, and changing perspective on how best to use the results of models in the regulatory process. The imperfect nature of modeling means that models will always have the potential for improvement through the integration of new scientific understandings and data sources. However, no advances in science, no matter how great, will ever make it possible to build a scientifically complete model or prove that a given model is correct in all respects. In addition, a more complete model is not necessarily a better one for the purposes of policy making. A good model is one that achieves the right balance between simplicity and complexity to address the question at hand.

The history of environmental analysis has focused on the primary need to understand the impacts of humans on the environment and to assess potential strategies to mitigate adverse impacts. This was the objective of *Man and Nature* (Marsh 1864) over 150 years ago—to describe "the character and approximately the extent of the changes produced by human actions in the physical conditions of the globe…" and to "suggest the possibility and the importance of the restoration of disturbed harmonies." Although the extent of the impacts and the models used to analyze impacts and develop responses look quite different today, these fundamental objectives remains the same. However, the successful use of new discoveries concerning environmental and human interactions is dependent on a holistic approach to generating data *and* interpreting the meaning of such data. Computational models will continue to provide linkages for interpretation, but as science gets more complex, it can easily become more isolated from nonscientists, whose distrust of science might increase. Ultimately, this can seriously damage the scientific endeavor. Thus, it is incumbent on both scientists and nonscientists to develop a strong communication bridge. Scientists need to find ways to express their findings to nonscientists. Nonscientists also have an obligation to seek more in-depth understanding of science. Finally, both scientists and nonscientists need to resist the temptation of wanting models to provide simple answers to the complex questions of the interrelationships of humans and the environment.

Epilogue

"Any philosophy that in its quest for certainty ignores the reality of the uncertain in the ongoing processes of nature denies the conditions out of which it arises."

John Dewey, *The Quest for Certainty*, 1929

References

ADEC (Alaska Department of Environmental Conservation). 2001. State Air Quality Control Plan, Vol. 2. Analysis of Problems, Control Actions, Section III.C: Fairbanks Transportation Control Program. Adapted July 27, 2001. Alaska Department of Environmental Conservation, Juneau, AK.

Air Improvement Resources, Inc. 1999. Office of the State Auditor Final Report: 1999 Audit of the Colorado AIR Program, November 22, 1999. Novi, MI: Air Improvement Resources, Inc.

Allen, T.H. 2004. Letter to Michael Leavitt, Administrator, Environmental Protection Agency, by Thomas, H. Allen, U.S. House of Representatives. May 13, 2004 [online]. Available: http://www.rapca.org/publications/Documents/Letter_from_Congressman_Allen_to_Leavitt.pdf [accessed Nov. 6, 2006].

Andersen, M.E., R.S.H. Yang, H.J. Clewell III, and M.B. Reddy. 2005. Introduction: A historical perspective of the development and applications of PBPK models. Pp. 1-18 in Physiologically Based Pharmacokinetic (PBPK) Modeling, M.B. Reddy, R.S.H. Yang, H.J. Clewell III, and M.E. Andersen, eds. Hoboken, NJ: Wiley-Interscience.

AQMWG (Air Quality Management Work Group). 2005. Recommendations to the Clean Air Act Advisory Committee. January 2005 [online]. Available: http://www.epa.gov/air/caaac/aqm/report1-17-05.pdf [accessed April 27, 2007].

Arrhenius, S. 1896. On the influence of carbonic acid in the air upon the temperature of the ground. Philos. Mag. J. Sci. 41:237-275 [online]. Available: http://members.lycos.nl/ErrenWijlens/co2/arrhenius.html [accessed Oct. 30, 2006].

Atkinson, R., and A.C. Lloyd. 1984. Evaluation of kinetic and mechanistic data for modeling of photochemical smog. J. Phys. Chem. Ref. Data 13(2):315-444.

Babendreier, J.E., and K.J. Castleton. 2005. Investigating uncertainty and sensitivity in integrated, multimedia environmental models: Tools for FRAMES-3MRA. Environ. Model. Softw. 20(8):1043-1055.

Bachman, J. 2006. Presentation on Proposal PMcoarse NAAQS. AAQTF Meeting Summary and Decisions, February 28-March 2, 2006, Bethesda, MD [online]. Available: http://www.airquality.nrcs.usda.gov/AAQTF/

Documents/Washington_Feb_2006/Draft_Minutes.pdf [accessed Nov. 6, 2006].

Barnwell, T.O., L.C. Brown, and R.C. Whittemore. 2004. Importance of field data in stream water quality modeling using QUAL2E-UNCAS. J. Environ. Eng. 130(6):643-647.

Bassetti, W.H., and M. Woodward. 2005. Epidemiology: Study Design and Data Analysis, 2nd Ed. Boca Raton, FL: Chapman & Hall/CRC.

Bates, J. 2004. Exploratory climate analysis tools for environmental satellite and weather radar data. Pp. 12-27 in Statistical Analysis of Massive Data Streams: Proceedings of a Workshop. Washington, DC: National Academy Press [online]. Available: http://www.research.att.com/~divesh/meetings/mpds2003/schedule/bates-talk.ppt [accessed Dec. 4, 2006].

Beck, M.B. 1987. Water quality modeling: A review of the analysis of uncertainty. Water Resour. Res. 23(8):1393-1442.

Beck, M.B., ed. 2002a. Environmental Foresight and Models: A Manifesto, 1st Ed. New York: Elsevier.

Beck, M.B. 2002b. Model evaluation and performance. Pp. 1275-1279 in Encyclopedia of Environmetrics, A.H. El-Shaarawi and W.W. Piegorsch, eds. Chichester, UK: Wiley.

Beck, M.B. 2005. Environmental foresight and structural change. Environ. Model. Softw. 20(6):651-670.

Beck, M.B., and J. Chen. 2000. Assuring the quality of models designed for predictive tasks. Pp. 401-420 in Sensitivity Analysis, A. Saltelli, K. Chan, and E.M. Scott, eds. Chichester, UK: Wiley.

Beck, M.B., J.R. Ravetz, L.A. Mulkey, and T.O. Barnwell. 1997. On the problem of model validation for predictive exposure assessments. Stoch. Hydrol. Hydraul. 11(3):229-254.

Beven, K.J. 1996. Equifinality and uncertainty in geomorphological modelling. Pp. 289-313 in The Scientific Nature of Geomorphology, B.L. Rhoades, and C.E. Thorn, eds. Chichester, UK: Wiley.

Beven, K.J. 2005. On the concept of model structure error. Water Sci. Technol. 52(6):167-175.

Bock, C.A. 1918. History of the Miami Flood Control Project, Part II. The Miami Conservancy District, Dayton, OH [online]. Available: http://cee.engr.ucdavis.edu/faculty/lund/Classes/MiamiConservancy.pdf [accessed Oct. 30, 2006].

Bredehoeft, J.D. 2003. From models to performance assessment: The conceptualization problem. Ground Water 41(5):571-577.

Bredehoeft, J.D. 2005. The conceptualization model problem-surprise. Hydrogeol. J. 13(1):37-46.

Bredehoeft, J.D., and L.F. Konikow. 1993. Ground-water models: Validate or invalidate. Ground Water 31(2):178-179.

Burden, D. 2004. Environmental Decision Making: Principles and Criteria for Groundwater Fate and Transport Models. Presentation at the First Meeting on Models in the Regulatory Decision Process, March 18, 2004, Washington, DC.

Burges, S.J. 1979. Water resources systems planning in USA: 1776-1976. J. Water Res. Pl.-ASCE 105(1):91-111.

Burtraw, D., and E. Mansur. 1999. Environmental effects of SO_2 trading and banking. Environ. Sci. Technol. 33(20):3489-3494.

CARB (California Air Resources Board). 1994. 1994 Low Emission Vehicle and Zero-Emission Vehicle Program Review: Staff Report. Sacramento, CA: California Environmental Protection Agency Air Resources Board. April 1994.

CARB (California Air Resources Board). 2000a. Evaluation of California's Enhanced Vehicle Inspection and Maintenance Program (Smog Check II). California Air Resources Board, Sacramento, CA.

CARB (California Air Resources Board). 2000b. Staff Report: 2000 Zero Emission Vehicle Program Biennial Review. California Environmental Protection Agency Air Resources Board. August 7, 2000 [online]. Available: http://www.arb.ca.gov/msprog/zevprog/2000review/staffreportfinal.pdf [accessed July 14, 2005].

CARB (California Air Resources Board). 2004. Final Statement of Reasons: The 2003 Amendments to the California Zero Emission Vehicle Program Regulations. California Environmental Protection Agency Air Resources Board. January 2004 [online]. Available: http://www.arb.ca.gov/regact/zev2003/fsor.pdf [accessed July 14, 2005].

Caswell, H. 1976. The validation problem. Pp. 313-325 in Systems Analysis and Simulation in Ecology, B.C. Patten, ed., Vol. IV. New York: Academic Press.

CCSP (Climate Change Science Program). 2003. Strategic Plan for Climate Change Science Program Final Report. July 2003 [online]. Available: http:/www.climatescience.gov/Library/stratplan2003/final/ccspstraplan2003-all.pdf [accessed Nov. 6, 2006].

Chameides, W.L., R.W. Lindsay, J. Richardson, and C.S. Kiang. 1988. The role of biogenic hydrocarbons in urban photochemical smog: Atlanta as a case study. Science 241(4872):1473-1475.

Chave, J., and S.A. Levin. 2003. Scale and scaling in ecological and economic systems. Environ. Resour. Econ. 26(4):527-557.

Chow, J.C., J.G. Watson, D. Crow, D.H. Lowenthal, and T. Merrifield. 2001. Comparison of IMPROVE and NIOSH carbon measurements. Aerosol Sci. Technol. 34(1):23-34.

Clarke, D. 2004. Models a la Mode – The Promise and Peril of Integrated Environmental Modeling. Foresight and Governance Project. Publication 2004-2. Woodrow Wilson International Center for Scholars, Washington, DC [online]. Available: http://www.environmentalfutures.org/Images/environmentalmodels.pdf [accessed Dec. 4, 2006].

Clemen, R.T. 1995. Making Hard Decisions: An Introduction to Decision Analysis. Belmont, CA: Duxbury Press.

Clewell, H. 2005. The Use of Physiologically Based Pharmacokinetic Modeling to Improve Chemical Dosimetry in Environmental Risk Assessments. Presentation at the Sixth Meeting on Models in the Regulatory Decision Process, December 2, 2005, Washington, DC.

CMAS (Community Modeling and Analysis System). 2006. CMAQ 4.6 Model [online]. Available: www.cmascenter.org/html/models.html [accessed Dec. 14, 2006].

Conolly, R. 2005. Computational Systems Biology: The Integration of Data Across Multiple Levels of Biological Organization to Understand How Perturbations of Normal Biology Become Adverse Health Effects. Presentation at the Sixth Meeting on Models in the Regulatory Decision Process, December 2, 2005, Washington, DC.

Crowther, J. 1924. Some considerations relative to the action of x-rays on tissue cells. Proc. Roy. Soc. Lond. B. Biol. Sci. 96(674):207-211.

Darier, E., C. Gough, B. De Marchi, S. Funtowicz, R. Grove-White, D. Kitchener, Â. Guimarães-Pereira, S. Shackley, and B. Wynne. 1999. Between democracy and expertise? Citizens' participation and environmental integrated assessment in Venice (Italy) and St. Helens (UK). J. Environ. Policy Plan. 1(2):103-120.

Daum, P., J. Meagher, D. Allen, and C. Durrenberger. 2002. Accelerated Science Evaluation of Ozone Formation in the Houston-Galveston Area, Executive Summary. Science Synthesis Committee. November 13, 2002 [online]. Available: http://www.utexas.edu/research/ceer/texaqsarchive/pdfs/EXEC_SUMMARY_Nov_02.pdf [accessed Oct. 30, 2006].

Dennis, R.L. 2002. The ozone problem. Pp. 147-169 in Environmental Foresight and Models: A Manifesto, M.B. Beck, ed. New York: Elsevier.

Dimitriades, B., ed. 1977. International Conference on Photochemical Oxidant Pollution and its Control: Proceedings, Vol. II. U.S. Environmental Protection Agency, Environmental Sciences Research Laboratory, Research Triangle Park, NC.

Diver, C.S. 1983. The optimal precision of administrative rules. Yale Law J. 93(1):65-109.

Dodge, M.C. 1977. Combined use of modeling techniques and smog chamber data to derive ozone-precursor relationships. Pp. 881-889 in International Conference on Photochemical Oxidant Pollution and Its Control, Vol. II, B. Dimitriades, ed. EPA/600/3-77-001a. Environmental Sciences Research Laboratory, U.S. Environmental Protection Agency, Research Triangle Park, NC.

Dolwick, P. 2005. Supplemental Analysis and Weight of Evidence. Presentation at the Third Particulate Matter/Regional Haze/Ozone Modeling Workshop, May 20, 2005, New Orleans, LA [online]. Available: http://cleanairinfo.com/modelingworkshop/presentations/SA_WOE_Dolwick.pdf [accessed Dec. 4, 2006].

Ducca, F. 2004. TRANSIMS – Public Private Cooperation in Software Development. Presentation at the Third Meeting on Models in the Regulatory Decision Process, December 2, 2004, Washington, DC.

Eagleton, K. 1999. Mathematical Techniques in Astronomy. Department of History and Philosophy of Science of the University of Cambridge [online]. Available: http://www.hps.cam.ac.uk/starry/mathematics.html [accessed Oct. 30, 2006].

EC (European Commission). 2004. Joint Workshop on Marine Assessment and Monitoring with Emphasis on Eutrophication, Istambul, 21-22 April 2004, Outcome of the Workshop. European Commission Joint Research Centre [online]. Available: http://sea.helcom.fi/dps/docs/documents/Monitoring %20and%20Assessment%20Group%20(MONAS)/MONAS%20Indicator %20Workshop%201,%202004/5-1.pdf [accessed Dec. 14, 2006].

Efroymson, R.E., and D.L. Murphy. 2001. Ecological risk assessment of multimedia hazardous air pollutants: Estimating exposure and effects. Sci. Total Environ. 274(1-3):219-230.

Ellerman, A.D., R. Schmalensee, P.L. Joskow, J.P. Montero, and E.M. Bailey. 1997. Emissions Trading under the U.S. Acid Rain Program: Evaluation of Compliance Costs and Allowance Market Performance. MIT Center for Energy and Environmental Policy Research, Cambridge, MA [online]. Available: http://web.mit.edu/ceepr/www/napap.pdf [accessed Nov. 6, 2006].

Engineering Staff of the Miami Conservancy District. 1922. Appraisal of flood protection benefits and damages. Engineering News-Record 89(Nov. 16):831-835.

EPA (U.S. Environmental Protection Agency). 1971. Air Quality Criteria for Nitrogen Oxides. AP-84. U.S. Environmental Protection Agency, Washington, DC. January.

EPA (U.S. Environmental Protection Agency). 1989. Resolution on the Use of Mathematical Models by EPA for Regulatory Assessment and Decision-Making. EPA-SAB-EEC-89-012. Environmental Engineering Committee, Science Advisory Board, U.S. Environmental Protection Agency. January 1989 [online]. Available: http://www.epa.gov/ord/crem/library/sab_89 resolution_models.pdf [accessed Nov. 7, 2006].

EPA (U.S. Environmental Protection Agency). 1992. Guidelines for Exposure Assessment. EPA/600/Z-92/001. Risk Assessment Forum, U.S. Environmental Protection Agency, Washington, DC. May 1992 [online]. Available: www.epa.gov/ncea/pdfs/guidline.pdf [accessed Feb. 27, 2007].

EPA (U.S. Environmental Protection Agency). 1994a. Report of the Agency Task Force on Environmental Regulatory Modeling: Guidance, Support Needs, Draft Criteria and Charter. EPA 500-R94-001. Office of Solid Waste and Emergency Response, U.S. Environmental Protection Agency, Washington, DC. March 1994.

EPA (U.S. Environmental Protection Agency). 1994b. Guidance Manual for the Integrated Exposure Uptake Biokinetic Model for Lead in Children.

EPA/540/R-93/081. OSWER9285.7-15-1. PB93-963510. Office of Solid Waste and Emergency Response, U.S. Environmental Protection Agency, Washington, DC. February 1994 [online]. Available: http://www.epa.gov/superfund/programs/lead/products.htm [accessed Nov. 2, 2006].

EPA (U.S. Environmental Protection Agency). 1994c. Agency Guidance for Conducting External Peer Review of Environmental Regulatory Modeling. Council for Regulatory Environmental Modeling, U.S. Environmental Protection Agency. July 1994 [online]. Available: http://cfpub.epa.gov/crem/modelpr.cfm [accessed Nov. 7, 2006].

EPA (U.S. Environmental Protection Agency). 1996. Review of National Ambient Air Quality Standards for Ozone. Assessment of Scientific and Technical Information OAQPS Staff Paper. EPA-452/R-96-007. Office of Air Quality Planning and Standards, U.S. Environmental Protection Agency, Research Triangle Park, NC.

EPA (U.S. Environmental Protection Agency). 1997a. Benefits and Costs of the Clean Air Act, 1970 to 1990. Final Report to U.S. Congress. EPA 410/R-97-002. Office of Air Quality Planning and Standards, U.S. Environmental Protection Agency, Research Triangle Park, NC [online]. Available: http://www.epa.gov/oar/sect812/contsetc.pdf [accessed Jan. 20, 2006].

EPA (U.S. Environmental Protection Agency). 1997b. Guiding Principles for Monte Carlo Analysis. EPA/630/R-97/001. Risk Assessment Forum, U.S. Environmental Protection Agency, Washington, DC. March 1997 [online]. Available: http://www.epa.gov/ncea/raf/montecar.pdf [accessed Nov. 21, 2006].

EPA (U.S. Environmental Protection Agency). 1998. Guidelines for Ecological Risk Assessment. EPA/630/R-95/002F. Risk Assessment Forum, U.S. Environmental Protection Agency. April 1998 [online]. Available: http://www.epa.gov/superfund/programs/nrd/era.htm [accessed Nov. 7, 2006].

EPA (U.S. Environmental Protection Agency). 1999a. Final Report to Congress on Benefits and Costs of the Clean Air Act, 1990 to 2010. EPA Report to Congress. EPA 410-R-99-001. Office of Air and Radiation, Office of Policy, U.S. Environmental Protection Agency, Washington, DC. November 1999 [online]. Available: http://epa.gov/oar/sect812/1990-2010/chap1130.pdf [accessed Nov. 7, 2006].

EPA (U.S. Environmental Protection Agency). 1999b. Science Algorithms of the EPA Models-3 Community Multiscale Air Quality [CMAQ] Modeling System. EPA/600/R-99/030. Office of Research and Development, U.S. Environmental Protection Agency, Washington, DC [online]. Available: http://www.epa.gov/asmdnerl/CMAQ/CMAQscienceDoc.html [accessed Dec. 14, 2006].

EPA (U.S. Environmental Protection Agency). 1999c. Description of the MOBILE Highway Vehicle Emissions Factor Model. Office of Mobile Sources, Ann Arbor, MI.

EPA (U.S. Environmental Protection Agency). 1999d. Draft Guidance on the Use of Models and Other Analyses in Attainment Demonstrations for the

8-Hour Ozone NAAQS. EPA-454/R-99-004. Office of Air Quality Planning and Standards, U.S. Environmental Protection Agency, Research Triangle Park, NC.

EPA (U.S. Environmental Protection Agency). 1999e. Total Risk Integrated Methodology: Status Report. EPA-453/R-99-010. Office of Air Quality, Planning and Standards, Research Triangle Park, NC [online]. Available: http://www.epa.gov/ttn/fera/trim_fate.html#historical [accessed Nov. 21, 2006].

EPA (U.S. Environmental Protection Agency). 2000. Benchmark Dose Technical Guidance Document. EPA/630/R-00/001. External Review Draft. Risk Assessment Forum, U.S. Environmental Protection Agency, Washington, DC [online]. Available: http://www.epa.gov/nceawww1/pdfs/bmds/BMD-External_10_13_2000.pdf [accessed June 12, 2007].

EPA (U.S. Environmental Protection Agency). 2001a. Comparison of Jeffords-Lieberman and Smith-Voinovich-Brownback. Office of Air and Radiation, U.S. Environmental Protection Agency [online]. Available: http://www.epa.gov/air/finalanalyses.pdf [accessed Oct. 30, 2006].

EPA (U.S. Environmental Protection Agency). 2001b. Economic Analysis of a Multi-Emissions Strategy. Prepared for Senators James M. Jeffords and Joseph I. Lieberman by Office of Air and Radiation, Office of Atmospheric Programs, U.S. Environmental Protection Agency, Washington, DC. October 31, 2001 [online]. Available: http://www. epa.gov/oar/jeffordslieberm.pdf [accessed Nov. 7, 2006].

EPA (U.S. Environmental Protection Agency). 2001c. General Principles for Performing Aggregate Exposure and Risk Assessments, Final Report. Office of Pesticide Programs, Office of Prevention, Pesticides, and Toxic Substances, U.S. Environmental Protection Agency, Washington, DC. December 2, 2001 [online]. Available: http://www.epa.gov/oppfead1/trac/science/ [accessed Nov. 7, 2006].

EPA (U.S. Environmental Protection Agency). 2001d. Better Assessment Science Integrating Point and Nonpoint Sources (BASINS), User's Manual, Version 3.0. EPA-823-H-01-001. Office of Water, U.S. Environmental Protection Agency, Washington, DC.

EPA (U.S. Environmental Protection Agency). 2002a. Guidelines for Ensuring and Maximizing the Quality, Objectivity, Utility, and Integrity of Information Disseminated by the Environmental Protection Agency. EPA/260R-02-008. Office of Environmental Information, U.S. Environmental Protection Agency, Washington, DC. October 2002 [online]. Available: http://www.epa.gov/QUALITY/informationguidelines/documents/EPA_InfoQualityGuidelines.pdf [accessed Nov. 1, 2006].

EPA (U.S. Environmental Protection Agency). 2002b. Total Maximum Daily Load for Total Mercury in the Ochlockonee Watershed, GA. Region 4, U.S. Environmental Protection Agency. February 28, 2002 [online]. Available: http://www.epa.gov/Region4/water/tmdl/georgia/ochlockonee/final_tmdls/OchlockoneeHgFinalTMDL.pdf [accessed Nov. 7, 2006].

EPA (U.S. Environmental Protection Agency). 2002c. Perchlorate Environmental Contamination: Toxicological Review and Risk Characterization. External Review Draft. NCEA-1-0503. National Center for Environmental Assessment, Office of Research and Development, U.S. Environmental Protection Agency, Washington, DC [online]. Available: http://cfpub.epa.gov/ncea/cfm/recordisplay. cfm?deid=24002 [accessed Nov. 7, 2006].

EPA (U.S. Environmental Protection Agency). 2002d. Guidance on Cumulative Risk Assessment of Pesticide Chemicals That Have a Common Mechanism of Toxicity. Office of Pesticide Programs, Office of Prevention, Pesticides, and Toxic Substances, U.S. Environmental Protection Agency, Washington, DC [online]. Available: http://www.epa.gov/oppfead1/trac/science/ [accessed Nov. 7, 2006].

EPA (U.S. Environmental Protection Agency). 2002e. Guidance for Quality Assurance Project Plans for Modeling. EPA/240/R-02/007. Office of Environmental Information, Environmental Protection Agency, Washington, DC. December 2002 [online]. Available: http://www.epa.gov/quality/qs-docs/g5m-final.pdf [accessed Nov. 22, 2006].

EPA (U.S. Environmental Protection Agency). 2003a. Council for Regulatory Environmental Modeling. Memorandum to Assistant Administrators, Associate Administrators, Regional Administrators and Science Policy Council, from Administrator Christine Todd Whitman, U.S. Environmental Protection Agency, Washington, DC. February 7, 2003 [online]. Available: http://www.epa.gov/ord/crem/library/whitman.PDF [accessed Nov. 1, 2006].

EPA (U.S. Environmental Protection Agency). 2003b. A Framework for a Computational Toxicology Research Program in ORD. EPA/600/R-03/065. Office of Research and Development, U.S. Environmental Protection Agency, Washington, DC. November 2003 [online]. Available: http://www.epa.gov/comptox/publications/comptoxframework06_02_04.pdf [accessed May 18, 2007].

EPA (U.S. Environmental Protection Agency). 2003c. RE: Request for Response to/Renewal of Federal Data Quality Act Petition Against Further Dissemination of "Climate Action Report 2002." Letter to Information Officer, Office of Environmental Information, U.S. Environmental Protection Agency, Washington, DC, from Christopher C. Horner, Competitive Enterprise Institute [online]. Available: http://www.epa.gov/quality/informationguidelines/documents/7428.pdf [accessed Nov. 8, 2006].

EPA (U.S. Environmental Protection Agency). 2003d. Draft Guidance on the Development, Evaluation, and Application of Regulatory Environmental Models. The Council for Regulatory Environmental Modeling (CREM), Office of Science Policy, Office of Research and Development, Washington, DC. November 2003 [online]. Available: http://www.epa.gov/ord/crem/library/CREM%20Guidance%20Draft%2012_03.pdf [accessed Nov. 21, 2006].

EPA (U.S. Environmental Protection Agency). 2003e. Multimedia, Multipathway, and Multireceptor Risk Assessment (3MRA) Modeling System, Vol. III. Ensuring Quality of the System, Modules, and Data. EPA 530-D-03-001c. SAB Review Draft. Prepared by U.S. Environmental Protection Agency, Office of Research and Development, National Exposure Research Laboratory, Athens, GA, and Research Triangle Park, NC; Office of Solid Waste, Washington, DC. July 2003 [online]. Available: http://www.epa.gov/epaoswer/haz waste/id/hwirwste/risk03.htm [accessed Nov. 21, 2006].

EPA (U.S. Environmental Protection Agency). 2003f. Users Guide to MOBILE6.1 and MOBILE6.2: Mobile Source Emission Factor. EPA420-R-03-010. Office of Transportation and Air Quality, U.S. Environmental Protection Agency, Ann Arbor, MI [online]. Available: http://www.epa.gov/OTAQ/models/mobile6/420r03010.pdf [accessed Nov. 16, 2006].

EPA (U.S. Environmental Protection Agency). 2003g. TRIM.FaTE User's Guide. Office of Air Quality Planning and Standards, U.S. Environmental Protection Agency. March 2003 [online]. Available: http://www.epa.gov/ttn/fera/trim_fate.html#current_user [accessed Nov. 4, 2006].

EPA (U.S. Environmental Protection Agency). 2003h. Multimedia, Multipathway, and Multireceptor Risk Assessment (3MRA) Modeling System, Vol. I. Modeling System and Science. EPA 530-D-03-001a. SAB Review Draft. Prepared by U.S. Environmental Protection Agency, Office of Research and Development, National Exposure Research Laboratory, Athens, GA, and Research Triangle Park, NC; Office of Solid Waste, Washington, DC. July 2003 [online]. Available: http://www.epa.gov/epaoswer/hazwaste/id/hwirwste/risk03.htm#vol1 [accessed Dec. 5, 2006].

EPA (U.S. Environmental Protection Agency). 2004a. An Examination of EPA Risk Assessment Principles and Practices. EPA/100/B-04/001. Office of the Science Advisor, U.S. Environmental Protection Agency, Washington, DC. March 2004 [online]. Available: http://www.epa.gov/OSA/pdfs/ratf-final.pdf [accessed Nov. 1, 2006].

EPA (U.S. Environmental Protection Agency). 2004b. Final Regulatory Impact Analysis: Control of Emissions from Nonroad Diesel Engines. EPA 420-R-04-007. Assessment and Standards Division, Office of Transportation and Air Quality, U.S. Environmental Protection Agency [online]. Available: http://www.epa.gov/nonroad-diesel/2004fr/420r04007a.pdf [accessed Nov. 9, 2006].

EPA (U.S. Environmental Protection Agency). 2004c. Air Quality Criteria for Particulate Matter, Vol. 1 and 2. EPA 600/P-99/002aF-bF. National Center for Environmental Assessment, Office of Research and Development, U.S. Environmental Protection Agency, Research Triangle Park, NC [online]. Available: http://cfpub.epa.gov/ncea/cfm/partmatt.cfm [accessed Nov. 9, 2006].

EPA (U.S. Environmental Protection Agency). 2004d. A Model Comparison: Dietary (Food and Water) Exposure in DEEM/Calendex, CARES, and

Lifeline. Office of Pesticide Programs, U.S. Environmental Protection Agency. Prepared for FIFRA Scientific Advisory Panel Meeting, April 28-29, 2004, Arlington, VA [online]. Available: http://www.epa.gov/scipoly/sap/meetings/2004/index.htm [accessed Nov. 9, 2006].

EPA (U.S. Environmental Protection Agency). 2004e. EPA's Multimedia, Multipathway, and Multireceptor Risk Assessment (3MRA) Modeling System; A Review by the 3MRA Review Panel of the EPA Science Advisory. EPA-SAB-05-003. EPA Science Advisory Board, U.S. Environmental Protection Agency. November 2004 [online]. Available: http://www.epa.gov/sab/pdf/sab_05_003.pdf [accessed Nov. 9, 2006].

EPA (U.S. Environmental Protection Agency). 2004f. Policy Guidance on the Use of MOBILE6.2 and the December 2003 AP-42 Method for Re-Entrained Road Dust for SIP Development and Transportation Conformity. Memorandum from Margo Oge, Director, Office of Transportation and Air Quality to EPA Regional Air Division Directors. February 24, 2004 [online]. Available: http://www.epa.gov/OTAQ/models/mobile6/mobil6.2_letter.pdf [accessed Nov. 16, 2006].

EPA (U.S. Environmental Protection Agency). 2004g. Potential Implications of Genomics for Regulatory and Risk Assessment Applications at EPA. EPA-100/B-04/002. Science Policy Council, U.S. Environmental Protection Agency, Washington, DC [online]. Available: http://www.epa.gov/osa/pdfs/EPA-Genomics-White-Paper.pdf [accessed Dec. 5, 2006].

EPA (U.S. Environmental Protection Agency). 2004h. Nationwide Identification of Hardrock Mining Sites. Evaluation report. Report No. 2004-P-00005. Office of Inspector General, U.S. Environmental Protection Agency, Washington, DC. March 31, 2004 [online]. Available: http://www.epa.gov/oig/reports/2004/20040331-2004-p-00005.pdf [accessed Dec. 5, 2006].

EPA (U.S. Environmental Protection Agency). 2005a. Guidelines for Carcinogen Risk Assessment EPA/630/P-03/001F. Risk Assessment Forum, U.S. Environmental Protection Agency, Washington, DC. March 2005 [online]. Available: http://cfpub.epa.gov/ncea/cfm/recordisplay.cfm?deid=116283 [accessed Nov. 9, 2006].

EPA (U.S. Environmental Protection Agency). 2005b. Review of EPA Region 5 Critical Ecosystem Assessment Model (CrEAM). EPA-SAB-05-011. Science Advisory Board, U.S. Environmental Protection Agency [online]. Available: http://www.epa.gov/sab/pdf/cream_sab-05-011.pdf [accessed Nov. 10, 2006].

EPA (U.S. Environmental Protection Agency). 2005c. A Set of Scientific Issues Being Considered by the Environmental Protection Agency Regarding: Preliminary N-Methyl Carbamate Cumulative Risk Assessment. SAP Minutes No. 2005-04. FIFRA Scientific Advisory Panel Meeting, August 23-26, 2005, Arlington, VA [online]. Available: http://www.epa.gov/scipoly/sap/meetings/2005/august/minutes.pdf [accessed Nov. 10, 2006].

EPA (U.S. Environmental Protection Agency). 2005d. Guidance on the Use of Models and Other Analyses in Attainment Demonstrations for the 8-hour Ozone NAAQS. Draft Final. Office of Air and Radiation, Office of Air Quality Planning and Standards, Research Triangle Park, NC. February 17, 2005 [online]. Available: http://www.epa.gov/ttn/scram/guidance/guide/draft-final-o3.pdf [accessed April 27, 2007].

EPA (U.S. Environmental Protection Agency). 2005e. Regulatory Impact Analysis of the Clean Air Mercury Rule: Final Report. EPA-452/R-05-003. Air Quality Strategies and Standards Division, Office of Air Quality Planning and Standards, U.S. Environmental Protection Agency [online]. Available: http://www.epa.gov/ttn/atw/utility/ria_final.pdf [accessed March 2, 2007].

EPA (U.S. Environmental Protection Agency). 2006a. Peer Review Handbook, 3rd Ed. EPA/100/B-06/002. Prepared for the U.S. Environmental Protection Agency by Members of the Per Review Advisory Group, for EPA's Science Policy Council, Washington, DC [online]. Available: http://epa.gov/peerreview/pdfs/Peer%20Review%20HandbookMay06.pdf [accessed Nov. 10, 2006].

EPA (U.S. Environmental Protection Agency). 2006b. Emission Standards for Hazardous Air Pollutants. Air Toxic Website, U.S. Environmental Protection Agency [online]. Available: http://www.epa.gov/ttn/atw/mactfnlalph.html [accessed Nov. 13, 2006].

EPA (U.S. Environmental Protection Agency). 2006c. Models Knowledge Base. Regulatory Environmental Modeling, U.S. Environmental Protection Agency [online]. Available: http://cfpub.epa.gov/crem/knowledge_base/knowbase.cfm [accessed Nov. 13, 2006].

EPA (U.S. Environmental Protection Agency). 2006d. Review of Agency Draft Guidance on the Development, Evaluation, and Application of Regulatory Environmental Models and Models Knowledge Base by the Regulatory Environmental Modeling Guidance Review Panel of the EPA Science Advisory Board. EPA-SAB-06-009. EPA Science Advisory Board, U.S. Environmental Protection Agency [online]. Available: http://www.epa.gov/science1/pdf/sab_06_009.pdf [accessed Nov. 10, 2006].

EPA (U.S. Environmental Protection Agency). 2006e. A Powerful Tool for Managing Watersheds. Better Assessment Science Integrating Point and Nonpoint Sources, U.S. Environmental Protection Agency [online]. Available: http://www.epa.gov/waterscience/BASINS/ [accessed Nov. 13, 2006].

EPA (U.S. Environmental Protection Agency). 2006f. Enforcement Economic Models. Civil Enforcement, U.S. Environmental Protection Agency [online]. Available: http://www.epa.gov/compliance/civil/econmodels/index.html [accessed Feb. 27, 2007].

EPA (U.S. Environmental Protection Agency). 2006g. EPA's Regulatory Impact Analyses (RIA) for the Revised Ozone and PM NAAQS and Proposed Regional Haze Rule. OAR Policy and Guidance, U.S. Environmental Protection Agency [online]. Available: http://www.epa.gov/ttn/oarpg/naaqsfin/ria.html [accessed Nov. 14, 2006].

EPA (U.S. Environmental Protection Agency). 2006h. Morbidity and Mortality: How Do We Value the Risk of Death and Illness? National Center for Environmental Economics, U.S. Environmental Protection Agency [online]. Available: http://yosemite.epa.gov/ee/epa/eerm.nsf/vwRepNumLookup/EE-0497?OpenDocument [accessed Nov. 13, 2006].

EPA (U.S. Environmental Protection Agency). 2006i. Information Quality Guidelines-Requests for Correction (RFC) and Requests for Reconsideration (RFR) Submitted to EPA. EPA Quality Guidelines, U.S. Environmental Protection Agency [online]. Available: http://www.epa.gov/quality/informationguidelines/iqg-list.html [accessed Nov. 14, 2006].

EPA (U.S. Environmental Protection Agency). 2006j. MOBILE6 Vehicle Emission Modeling Software. Office of Transportation and Air Quality, U.S. Environmental Protection Agency [online]. Available: http://www.epa.gov/oms/m6.htm [accessed Dec. 14, 2006].

EPA (U.S. Environmental Protection Agency). 2006k. TRIM.FaTE Evaluation Plan Follow-up. U.S. Environmental Protection Agency [online]. Available: http://www.epa.gov/ttn/fera/trim_fate.html [accessed Nov. 21, 2006].

EPA (U.S. Environmental Protection Agency). 2006l. Modeling Applications and Tools. Support Center for Regulatory Atmospheric Modeling. EPA Technology Transfer Network, U.S. Environmental Protection Agency [online]. Available: http://www.epa.gov/scram001/dispersion_prefrec.htm [accessed Nov. 30, 2006].

EPA (U.S. Environmental Protection Agency). 2006m. Preferred/Recommended Models. Support Center for Regulatory Atmospheric Modeling. EPA Technology Transfer Network, U.S. Environmental Protection Agency [online]. Available: http://www.epa.gov/scram001/dispersion_prefrec.htm [accessed Nov. 30, 2006].

EPA (U.S. Environmental Protection Agency). 2006n. Guidance on the Use of Models and Other Analyses for Demonstrating Attainment of Air Quality Goals for Ozone, PM2.5, and Regional Haze, Office of Air Quality Planning and Standards, Air Quality Modeling Group, U.S. Environmental Protection Agency, Research Triangle Park, NC. September 2006 [online]. Available: http://www.epa.gov/scram001/guidance/guide/draft_final-pm-O3-RH.pdf [accessed Dec. 5, 2006].

EPA (U.S. Environmental Protection Agency). 2007. Enforcement Economic Models. Office of Civil Enforcement, U.S. Environmental Protection Agency [online]. Available: http://www.epa.gov/Compliance/civil/econmodels [accessed June 12, 2007].

Fenner, K., M. Scheringer, M. MacLeod, M. Matthies, T. McKone, M. Stroebe, A. Beyer, M. Bonnell, A.C. Le Gall, J. Klasmeier, D. Mackay, D. van de Meent, D. Pennington, B. Scharenberg, N. Suzuki, and F. Wania. 2005.

Comparing estimates of persistence and long-range transport potential among multimedia models. Environ. Sci. Technol. 39(7):1932-1942.

Finkel, A.M. 1990. Confronting Uncertainty in Risk Management: A Guide for Decision Makers. Washington, DC: Resources for the Future.

Finkel, A.M. 2004. Guidelines for Choosing Models that Encourage Research, Reveal the Inevitable Value Judgments, and Promote Sensible Decisions. Presentation at the First Meeting on Models in the Regulatory Decision Process, March 18, 2004, Washington, DC.

Finkel, A.M., and J.S. Evans. 1987. Evaluating the benefits of uncertainty reduction in environmental health risk management. JAPCA 37(10):1164-1171.

Fluent, Inc. 2006. A Brief History of Computational Fluid Dynamics [online]. Available: http://www.fluent.com/about/cfdhistory.htm [accessed Nov. 2, 2006].

Foley, G. 2004. U.S. EPA's Council for Regulatory Environmental Modeling (CREM). Presentation at the First Meeting on Models in the Regulatory Decision Process, March 18, 2004, Washington, DC.

Foley, G. 2005. Global Earth Observation System of Systems (GEOSS) and Environmental Regulatory Modeling. Presentation at the Sixth Meeting on Models in the Regulatory Decision Process, December 2, 2005, Washington, DC.

Foley, G.J., P.G. Georgopoulos, and P.J. Lioy. 2003. Accountability within new ozone standards. Environ. Sci. Technol. 37(21):392A-399A.

Freeman, A.M. 1978. Air and water pollution policy. Pp. 12-67 in Current Issues in U.S. Environmental Policy, P. Portney, ed. Baltimore: Johns Hopkins University Press.

Frey, C. 2004. Methods and Applications of Uncertainty and Sensitivity Analysis for Models Used in Regulatory Processes. Presentation at the Third Meeting on Environmental Decision Making: Principles and Criteria for Models, December 2, 2004, Washington, DC.

Fujita, E.M., B.E. Croes, C.L. Bennett, D.R. Lawson, F.W. Lurmann, and H.H. Main. 1992. Comparison of emission inventory and ambient concentration ratios of CO, NMOG, and NOx in California's South Coast Air Basin. J. Air Waste Manage. Assoc. 42(3):264-276.

Funtowicz, S.O., and J.R. Ravetz. 1990. Uncertainty and Quality in Science for Policy. Dordrecht, The Netherlands: Kluwer.

GAO (U.S. General Accounting Office). 1996. Peer Review EPA's Implementation Remains Uneven: Report to Congressional Requesters. GAO/RCED-96-236. Washington, DC: U.S. General Accounting Office.

Gayer, T., and R.W. Hahn. 2005. Regulating mercury: What's at stake? Science 310(5749):777-779.

George, J. 2004. State Perspective on Modeling in Support of the TMDL Program. Presentation at the First Meeting on Models in the Regulatory Decision Process, March 18, 2004, Washington, DC.

Georgopoulos, P.G., S.W. Wang, V.M. Vyas, Q. Sun, A. Chandrasekar, P. Shade, R. Vedantham, J. Burke, T. McCurdy, and H. Özkaynak. 2005a. A

Source-to-Dose Assessment of Population Exposures to Fine PM, Ozone and Air Toxics in Philadelphia, PA, During a 1999 Summer Episode. A Case Study Employing MENTOR/SHEDS-1A. Technical Report CERM 2005-02. Computational Chemodynamics Laboratory, Environmental and Occupational Health Science Institute, and National Exposure Research Laboratory, U.S. Environmental Protection Agency. February 2005 [online]. Available: http://ccl.rutgers.edu/reports/cerm/CCL-CERM_Report-2005-01_philadelphia.pdf [accessed Dec. 5, 2006].

Georgopoulos, P.G., S.W. Wang, V.M. Vyas, Q. Sun, J. Burke, R. Vedantham, T. McCurdy, and H. Özkaynak. 2005b. A source-to-dose assessment of population exposures to fine PM and ozone in Philadelphia, PA, during a summer 1999 episode. J. Expo. Anal. Env. Epid. 15(5):439-457.

Gertler, A.W., J.C. Sagebiel, D.N. Wittorff, W.R. Pierson, W.A. Dippel, D. Freeman, and L. Sheetz. 1997. Vehicle Emissions in Five Urban Tunnels. NTIS/PB97-158091. Desert Research Institute, Nevada University System, Reno.

Gilliland, A. 2003. Overview of Model Evaluation Plans for CMAQ FY04 Release. Presented at the CMAQ Model Peer Review Meeting, December 17, 2003, Research Triangle Park, NC [online]. Available: http://www.cmascenter.org/r_and_d/first_review/pdf/model_evaluation_plans_for_cmaq04_(gilliland).pdf?temp_id=99999 [accessed Nov. 22, 2006].

Gilman, P. 2005. Computation in the Regulatory Environment: Solution or the Problem. Presentation at the Sixth Meeting on Models in the Regulatory Decision Process, December 2, 2005, Washington, DC.

Graham, J.D. 2004. OMB's Circular No. A-4, New Guidelines for the Conduct of Regulatory Analysis. Memorandum for the President's Management Council, from John Graham, Administrator, Office of Information and Regulatory Affairs, Office of Management and Budget, Executive Office of the President. March 2, 2004 [online]. Available: http://www.whitehouse.gov/omb/inforeg/memo_pmc_a4.pdf [accessed Nov. 10, 2006].

Greene, D.L., J. Kahn, and R. Gibson. 1999. Fuel economy rebound effect for U.S. households vehicles. Energy J. 20(3):1-31.

Gruenspecht, H.K. 1982. Differentiated regulation: The case of auto emissions standards. Am. Econ. Rev. 72(2):328-331.

Gustavson, K.E., L.W. Barnthouse, C.L. Brierley, E.H. Clarke, C.H. Ward. 2007. Superfund and mining megasites. Environ. Sci. Technol. 41(8): 2667-2672.

Hammitt, J.K. 1990. Subjective-probability-based scenarios for uncertain input parameters. Risk Anal. 10(1):93-102.

Harrington, W. 1989. Enforcement leverage when penalties are restricted. J. Public Econ. 37(1):29-53.

Harrington, W. 1997. Fuel economy and motor vehicle emissions. J. Environ. Econ. Manage. 33(3):240-252.

Henderson, F.M., and A.J. Lewis. 1998. Principles and Applications of Imaging Radar. New York: Wiley.
Herrick, C. 2000. Predictive modeling of acid rain: Obstacles to generating useful information. Pp. 251-268 in Prediction: Science, Decision Making and the Future of Nature, D. Sarewitz, R.A. Pielke, Jr., and R. Byerly, Jr., eds. Washington, DC: Island Press.
Herrmann, R.A. 1997. The beginnings of differential equation modeling. Pp. 76-87 in Nonstandard Analysis Applied to Advanced Undergraduate Mathematics: Infinitesimal Modeling and Very Elementary Physics. Annapolis, MD: U.S. Naval Academy [online]. Available: http://www.citebase.org/fulltext?format=application%2Fpdf&identifier=oai%3AarXiv.org%3Amath%2F0312432 [accessed Nov. 1, 2006].
Ho, C. 2004. Estimating Motor Vehicle Emissions: A Tale of Two Models. Presentation at the First Meeting on Models in the Regulatory Decision Process, March 19, 2004, Washington, DC.
Hoeting, J.A., D. Madigan, A. Raftery, and C.T. Volinsky. 1999. Bayesian model averaging: A tutorial. Stat. Sci. 14(4):382-417.
Holmes, K.J., and R.J. Cicerone. 2002. The road ahead for vehicle emission inspection and maintenance programs. EM(July):15-22.
Holmes, K.J., and J.H. Ellis. 1999. An integrated assessment modeling framework for assessing primary and secondary impacts from carbon dioxide stabilization scenarios. Environ. Model. Assess. 4(1):45-63.
Holmes, K.J., and A.G. Russell. 2001. Improving mobile-source emissions modeling. EM (February):20-28.
Holmes, K.J., and M.W. Wolman. 2001. The early development of systems analysis in natural resources management from man and nature to the Miami Conservancy District. Environ. Manage. 27(2):177-193.
Houk, I.E. 1921. Rainfall and Runoff in the Miami Valley. Technical Report Part 8. Dayton, OH: State of Ohio, Miami Conservancy District [online]. Available: http://www.uamv.com/documents/uam-v_1.31_user's_guide.pdf [accessed March 31, 2007].
Hubbard, T.N. 1998. An empirical examination of moral hazard in the vehicle inspection market. RAND J. Econ. 29(2):406-426.
Jacobson, M.Z. 2001. GATOR-GCMM: A global-through urban-scale air pollution and weather forecast model. 1. Model design and treatment of subgrid soil, vegetation, roads, rooftops, water, sea ice, and snow. J. Geophys. Res. 106(D6):5385-5402.
Jakeman, A.J., and G.M. Hornberger. 1993. How much complexity is warranted in a rainfall-runoff model? Water Resour. Res. 29(8):2637-2649.
Jakeman, A.J., R.A. Letcher, and J.P. Norton. 2006. Ten iterative steps in development and evaluation of environmental models. Environ. Modell. Softw. 21(5):602-614.
Janssen, M., and J. Rotmans. 1995. The Evolution of Integrated Assessment Models for Global Climate Change: Towards a New Generation. Prepared

for the Energy Modeling Forum Working Group Meeting, 24-26 May 1995, Stanford University, California.

Janssen, P.H.M., A.C. Petersen, J.P. van der Sluijs, J.S. Risbey, and J.R. Ravetz. 2005. A guidance for assessing and communicating uncertainties. Water Sci. Technol. 52(6):125-131.

Jasanoff, S. 1990. The Fifth Branch: Science Advisors as Policymakers. Cambridge, MA: Harvard University Press.

Jasanoff, S. 2004. Peer Review of Regulatory Models – The Hard Look and the Long View. Presentation at the Third Meeting on Environmental Decision Making: Principles and Criteria for Models, December 2, 2004, Washington, DC.

Jaworska, J.S, M. Comber, C. Auer, and C.J. Van Leeuwen. 2003. Summary of a workshop on regulatory acceptance of (Q)SARs for human health and environmental endpoints. Environ. Health Perspect. 111(10):1358-1360.

Kahn, R. 1997. Earth observation systems: What shall we do with the data we are expecting in 1998? Pp. 15-22 in Massive Data Sets: Proceedings of Workshop. Washington, DC: National Academy Press.

Kelman, S. 1982. What Price Incentives? Economists and the Environment. Westport, CT: Auburn House.

Klasmeier, J., M. Matthies, M. MacLeod, K. Fenner, M. Scheringer, M. Stroebe, A.C. Le Gall, T.E. McKone, D. van de Meent, and F. Wania. 2006. Application of multimedia models for screening assessment of long-range transport potential and overall persistence. Environ. Sci. Technol. 40(1):53-60.

Kleit, A.N. 1990. The effect of annual changes in automobile fuel economy standards. J. Regul. Econ. 2(2):151-172.

Kneese, A., and C. Schultze. 1975. Pollution, Prices, and Public Policy. Washington, DC: Brookings Institution.

Konikow, L.F., and J.D. Bredehoeft. 1992. Ground-water models cannot be validated. Adv. Water Resour. 15(1):75-83.

Koop, G., and L. Tole. 2004. Measuring the health effects of air pollution: To what extent can we really say that people are dying from bad air? J. Environ. Econ. Manage. 47(1):30-54.

Krupnick, A., R. Morgenstern, M. Batz, P. Nelsen, D. Butraw, J.S. Shih, and M. McWilliams. 2006. Not a Sure Thing: Making Regulatory Choices Under Uncertainty. Washington, DC: Resources for the Future. February 2006 [online]. Available: http://www.rff.org/rff/Documents/RFF-Rpt-RegulatoryChoices.pdf [accessed Nov. 22, 2006].

Kwoka, J.E., Jr. 1983. The limits of market-oriented regulatory techniques: The case of automotive fuel economy. Q. J. Econ. 98(4):695-704.

Lawson, D.R. 1993. Passing the test – human behavior and California's Smog Check Program. J. Air Waste Manage. Assoc. 43(12):1567-1575.

Lawson, D.R. 2003. The weekend ozone effect – The Weekly Ambient Emissions Control Experiment. EM (July):17-26.

Leighton, P.A. 1961. Photochemistry of Air Pollution. New York: Academic Press.

Liao, H., P.J. Adams, S.H. Chung, J.H. Seinfeld, L.J. Mickley, and D.J. Jacob. 2003. Interactions between tropospheric chemistry and aerosols in a unified general circulation model. J. Geophys. Res. 108(D1):4001 doi:10.1029/2001JD001260.

Liao, H., J.H. Seinfeld, P.J. Adams, and L.J. Mickley. 2004. Global radiative forcing of coupled tropospheric ozone and aerosols in a unified general circulation model. J. Geophys. Res. 109(D16):16207 doi:10.1029/2003JD004456.

Linkov, I., and D. Burmistrov. 2003. Model uncertainty and choices made by modelers: Lessons learned from the International Atomic Energy Agency Model Intercomparisons. Risk Anal. 23(6):1297-1308.

Lioy, P.J. 1990. Assessing total human exposure to contaminants. Environ. Sci. Technol. 24(7):938-945.

Lotka, A.J. 1925. Elements of Physical Biology. Baltimore: Williams and Wilkins Co.

Lund, J.R., and R.N. Palmer. 1997. Water resources system modeling for conflict resolution. Water Resour. Update 3(108):70-82 [online]. Available: http://www.tag.washington.edu/papers/papers/Lund-and-Palmer.1997.WaterResourcesUpdate.108,70-82.pdf [accessed Dec. 5, 2006].

Magat, W., A.J. Krupnick, and W. Harrington. 1986. Rules in the Making: A Statistical Analysis of Regulatory Agency Behavior. Washington, DC: Resources for the Future.

Mahoney, M.S. 1998. The mathematical realm of nature. Pp. 702-758 in Cambridge History of Seventeenth-Century Philosophy, Vol. 1, D.E. Garber and M. Ayers, eds. Cambridge, UK: Cambridge University Press.

Marsh, G.P. 1864. Man and Nature. Anotated reprint, 1965, Cambridge, MA: Harvard University Press.

McGarity, T.O., and W.E. Wagner. 2003. Legal aspects of the regulatory use of environmental modeling. Environ. Law Rep. 33(10):10751-10774.

Mesterton-Gibbons, M. 1995. A Concrete Approach to Mathematical Modelling. New York: J. Wiley & Sons.

Mollan, C. 2004. Orrery by Gilkerson and Co. National Inventory of Scientific Instruments, Royal Dublin Society. Armagh Observatory [online]. Available: http://star.arm.ac.uk/history/instruments/Glikerson-orrery.html [accessed Nov. 2, 2006].

Morales, K.H., L. Ryan, T.L. Kuo, M.M. Wu, and C.J. Chen. 2000. Risk of internal cancers from arsenic in drinking water. Environ. Health Perspect. 108(7):655-661.

Morales, K.H., J.G. Ibrahim, C.J. Chen, and L.M. Ryan. 2006. Bayesian model averaging with applications to benchmark dose estimation for arsenic in drinking water. J. Am. Stat. Assoc. 101(473):9-17.

Morgan, A.E. 1951. The Miami Conservancy District, 1st Ed. New York: McGraw-Hill.

Morgan, G. 2004. Parameter and Model Uncertainty in Models for Regulatory Decision Making: Problems and Opportunities. Presentation at the Third Meeting on Models in the Regulatory Decision Process, December 2, 2004, Washington, DC.

Morgan, G.M., and M. Henrion. 1990. Uncertainty: A Guide to Dealing with Uncertainty on Quantitative Risk and Policy Analysis. Cambridge, UK: Cambridge University Press.

Morgan, M.G., S.C. Morris, A.K. Meier, and D.L. Shenk. 1978. A probabilistic methodology for estimating air pollution health effects from coal-fired power plants. Energy Syst. Policy 2(3):287-310.

Morgan, M.G., S.C. Morris, M. Henrion, D.A.L. Amaral, and W.R. Rish. 1984. Technical uncertainty in quantitative policy analysis: A sulfur air pollution example. Risk Anal. 4(3):201-216.

Morgenstern, R. 2005. Making Regulatory Choices: What's Uncertainty Have to Do with It? Presentation at the Sixth Meeting on Models in the Regulatory Decision Process, December 2, 2005, Washington, DC.

Murphy, D.L., and H. Richmond. 2004. OAQPS Risk Assessment. Presentation to EPA Science Advisory Board, Joint Meeting of the Environmental Health Committee and Integrated Human Exposure Committee, July 26, 2004, Washington, DC.

Murrell, P. 2006. R Graphics. Boca Raton, FL: Chapman & Hall/CRC.

NAPAP (National Acid Precipitation Assessment Program). 1990. 1990 Integrated Assessment Report. Washington, DC: NAPAP Office of the Director.

Napolitano, S., and E. Lieberman. 2004. EPA Use of ICF's Integrated Planning Model. Presentation at the Third Meeting on Models in the Regulatory Decision Process, December 2, 2004, Washington, DC.

NARSTO. 2004. Particulate Matter Assessment for Policy Makers: A NARSTO Assessment, P. McMurry, M. Shepherd, and J. Vickery, eds. Cambridge, England: Cambridge University Press.

NARSTO. 2005. Improving Emissions Inventories for Effective Air Quality Management Across North America: A NARSTO Assessment. NARSTO-05-001. Prepared by the NARSTO Emission Inventory Assessment Team [online]. Available: http://www.narsto.org/section.src?SID=8 [accessed March 5, 2007].

Nowotny, H., P. Scott, and M. Gibbons. 2001. Re-Thinking Science: Knowledge and the Public in an Age of Uncertainty. Malden, MA: Polity/Blackwell.

NRC (National Research Council). 1983. Risk Assessment in the Federal Government: Managing the Process. Washington, DC: National Academy Press.

NRC (National Research Council). 1991. Rethinking the Ozone Problem in Urban and Regional Air Pollution. Washington, DC: National Academy Press.

NRC (National Research Council). 1994. Science and Judgment in Risk Assessment. Washington, DC: National Academy Press.

NRC (National Research Council). 1998. Research Priorities for Airborne Particulate Matter: I. Immediate Priorities and a Long-Range Research Portfolio. Washington, DC: National Academy Press.

NRC (National Research Council). 1999a. Arsenic in Drinking Water. Washington, DC: National Academy Press.

NRC (National Research Council). 1999b. Research Priorities for Airborne Particulate Matter: II. Evaluating Research Progress and Updating the Portfolio. Washington, DC: National Academy Press.

NRC (National Research Council). 2000. Modeling Mobile-Source Emissions. Washington, DC: National Academy Press.

NRC (National Research Council). 2001a. Evaluating Vehicle Emissions Inspection and Maintenance Programs. Washington, DC: National Academy Press.

NRC (National Research Council). 2001b. Arsenic in Drinking Water 2001 Update. Washington, DC: National Academy Press.

NRC (National Research Council). 2001c. Assessing the TMDL Approach to Water Quality Management. Washington, DC: National Academy Press.

NRC (National Research Council). 2001d. Research Priorities for Airborne Particulate Matter: III. Early Research Progress. Washington, DC: National Academy Press.

NRC (National Research Council). 2002. Estimating the Public Health Benefits of Proposed Air Pollution Regulations. Washington, DC: National Academies Press.

NRC (National Research Council). 2003. Adaptive Monitoring and Assessment for the Comprehensive Everglades Restoration Plan. Washington, DC: National Academies Press.

NRC (National Research Council). 2004a. Air Quality Management in the United States. Washington, DC: National Academies Press.

NRC (National Research Council). 2004b. Implementing Climate and Global Change Research: A Review of the Final U.S. Climate Change Science Program Strategic Plan. Washington, DC: National Academies Press.

NRC (National Research Council). 2004c. Research Priorities for Airborne Particulate Matter: IV. Continuing Research Progress. Washington, DC: National Academies Press.

NRC (National Research Council). 2005a. Superfund and Mining Megasites. Washington, DC: National Academies Press.

NRC (National Research Council). 2005b. Health Implications of Perchlorate Ingestion. Washington, DC: National Academies Press.

NRC (National Research Council). 2005c. State and Federal Standards for Mobile Source Emissions. Washington, DC: National Academies Press.

NRC (National Research Council). 2006a. Human Biomonitoring for Environmental Chemicals. Washington, DC: National Academies Press.

NRC (National Research Council). 2006b. Toxicity Testing for Assessment of Environmental Agents. Washington, DC: National Academies Press.

O'Connor, J.J., and E.F. Robertson. 2003. History Topic: Mathematics and the Physical World [online]. Available: http://www-groups.dcs.st-and.ac.uk/~history/HistTopics/World.html [accessed Nov. 2, 2006].

OECD (Organisation for Economic Co-operation and Development). 2002. Report of the OECD/UNEP Workshop on the Use of Multimedia Models for Estimating Overall Environmental Persistence and Long-Range Transport Potential in the Context of PBTs/POPs Assessment. ENV/JM/MONO (2002)15. OECD Series on Testing and Assessment No. 36. Organisation for Economic Co-operation and Development, Paris. August 9, 2002 [online]. Available: http://www.olis.oecd.org/olis/2002doc.nsf/43bb6130e5e86e5fc12569fa005d004c/150147753d5c7f6cc1256c010047d31d/$FILE/JT00130274.PDF [accessed April 27, 2007].

OECD (Organisation for Economic Co-operation and Development). 2004. Guidance Document on the Use of Multimedia Models for Estimating Overall Environmental Persistence and Long-Range Transport. ENV/JM/MONO(2004)5. OECD Series on Testing and Assessment No. 45. Organisation for Economic Co-operation and Development, Paris. March 18, 2004 [online]. Available: http://appli1.oecd.org/olis/2004doc.nsf/43bb6130e5e86e5fc12569fa005d004c/d5d405dd127503abc1256e5b00541567/$FILE/JT00160339.PDF [accessed April 27, 2007].

OMB (U.S. Office of Management and Budget). 2001. Guidelines for Ensuring and Maximizing the Quality, Objectivity, Utility, and Integrity of Information Disseminated by Federal Agencies. Washington, DC: Office of Management and Budget, Executive Office of the President [online]. Available: http://www.whitehouse.gov/omb/fedreg/final_information_quality_guidelines.html [accessed Oct. 12, 2006].

OMB (U.S. Office of Management and Budget). 2003. Regulatory Analysis. Circular A-4 to the Heads of Executive Agencies and Establishments, September 17, 2003 [online]. Available: http://www.whitehouse.gov/omb/circulars/a004/a-4.pdf [accessed Nov. 3, 2006].

OMB (U.S. Office of Management and Budget). 2004. Final Information Quality Bulletin for Peer Review. Office of Management and Budget. December 15, 2004 [online]. Available: http://www.whitehouse.gov/omb/inforeg/peer2004/peer_bulletin.pdf [accessed Oct. 12, 2006].

OMB (U.S. Office of Management and Budget). 2005. Validating Regulatory Analysis: 2005 Report to Congress on the Costs and Benefits of Federal Regulations and Unfunded Mandates on State, Local, and Tribal Entities. Office of Information and Regulatory Affairs, Office of Management and Budget, Washington, DC [online]. Available: http://www.whitehouse.gov/omb/inforeg/2005_cb/final_2005_cb_report.pdf [accessed Nov. 13, 2006].

Oreskes, N. 1998. Evaluation (not validation) of quantitative models. Environ. Health Perspect. 106(suppl. 6):1453-1460.

Oreskes, N., K. Shrader-Frechette, and K. Belitz. 1994. Verification, validation, and confirmation of numerical models in the earth sciences. Science 263(5147):641-646.

Pascual, G. 2004. Wresting Regulatory Decisions from an Uncertain World. Presentation at the Third Meeting on Models in the Regulatory Decision Process, December 2, 2004, Washington, DC.

Perciasepe, R. 2005. Use of Modeling in Decision-Making at EPA. Presentation at the Sixth Meeting on Models in the Regulatory Decision Process, December 2, 2005, Washington, DC.

Petersen, B. 2004. Why Use Proprietary Software for Risk Assessment. Presentation at the Third Meeting on Models in the Regulatory Decision Process, December 2, 2004, Washington, DC.

Pierce, R.G. 1988. Two problems in administrative law: Political polarity on the District of Columbia circuit and judicial deterrence of agency rulemaking. Duke Law J. 1988(2/3):300-328.

Pope, C.A. 1989. Respiratory disease associated with community air pollution and a steel mill, Utah Valley. Am. J. Public Health 79(5):623-628.

Portney, P.R., I.W.H. Parry, H.K. Gruenspecht, and W. Harrington. 2003. The Economics of Fuel Economy Standards. Paper No. 03-44. Washington, DC: Resources for the Future.

Preuss, P. 2006. EPA's Advances in Human Health Assessment Research. Presentation at the Joint Meeting of the Environmental Health Committee and the Integrated Human Exposure Committee, EPA Science Advisory Board, September 6, 2006, Washington, DC.

Ramaswami, A., J.B. Milford, and M.J. Small. 2005. Integrated Environmental Modeling – Pollutant Transport, Fate and Risk in the Environment. Hoboken, NJ: J. Wiley.

Reckhow, K. 2005. Integrating Modeling and Monitoring in Adaptive Strategies to Attain Water Quality Standards. Presentation at the Sixth Meeting on Models in the Regulatory Decision Process, December 2, 2005, Washington, DC.

Reddy, M.B, R.S. Yang, H.J. Clewell, and M.E. Andersen. 2005. Physiologically Based Pharmacokinetic Modeling: Science and Applications. Hoboken, NJ: Wiley.

Refsgaard, J.C., J.P. van der Sluijs, J. Brown, and P. van der Keur. 2006. A framework for dealing with uncertainty due to model structure error. Adv. Water Resour. 29(11):1586-1597.

Reynolds, S.D., P.M. Roth, and J.H. Seinfeld. 1973. Mathematical modeling of photochemical air pollution. 1. Formulation of the model. Atmos. Environ. 7:1033-1061.

Risbey, J., J.P. van der Sluijs, P. Kloprogge, J.R. Ravetz, S. Funtowicz, and S.C. Quintana. 2005. Application of a checklist for quality assistance in environmental modelling to an energy model. Environ. Model. Assess. 10(1):63-79.

Roberts, L. 1991. Learning from an acid rain program. Science 251(4999):1302-1305.

Rotmans, J. 1990. IMAGE: An Integrated Model to Assess the Greenhouse Effect. Dordrecht, The Netherlands: Kluwer.

Rubin, E.S. 1992. Global Warming Research: Learning from NAPAP's Mistakes. Contemporary Issues Series 53. Center for the Study of American Business, Washington University, St. Louis, MO [online]. Available: http://www.iecm-online.com/publications.html [accessed Dec. 11, 2005].

Rubin, K. 1985. Efficient Investment in Wastewater Treatment Plants. Washington, DC: Congressional Budget Office.

Russell, A., and R. Dennis. 2000. NARSTO critical review of photochemical models and modeling. Atmos. Environ. 34(12-14):2283-2324.

Sass, J. 2004. Natural Resource Defense Council's Perspective on Environmental Modeling. Presentation at the First Meeting on Models in the Regulatory Decision Process, March 18, 2004, Washington, DC.

Scheffe, R.D., and R.E. Morris. 1993. A review of the development and application of the Urban Airshed model. Atmos. Environ. B-Urb. 27(1):23-39.

Scheringer, M. 1996. Persistence and spatial range as endpoints of an exposure-based assessment of organic chemicals. Environ. Sci. Technol. 30(5):1652-1659.

Scheringer, M. 1997. Characterization of the environmental distribution behavior of organic chemicals by means of persistence and spatial range. Environ. Sci. Technol. 31(10):2891-2897.

Scheringer, M., M. MacLeod, and F. Wegmann. 2006. The OECD Pov and LRTP Screening Tool, Version 2.0 — Software and Manual. http://www.sust-chem.ethz.ch/downloads/Tool2_0_Manual.pdf [accessed April 27, 2007].

Schertzer, W.M., and D.C.L. Lam. 2002. Lake Erie and evolving issues of the quality of its water. Pp. 105-130 in Environmental Foresight and Models: A Manifesto, M.B. Beck, ed. Oxford: Elsevier.

Schichl, H. 2004. Models and history of modeling. Pp. 25-36 in Modeling Languages in Mathematical Optimization, J. Kallrath, ed. Boston: Kluwer.

Schnoor, J.L. 1996. Environmental Modeling: Fate and Transport of Pollutants in Water, Air, and Soil. New York: Wiley.

Schwab, M. 2004. OMB Experience (and Frustration) with Models Used in Environmental Decision Making. Presentation at the Third Meeting on Models in the Regulatory Decision Process, December 2, 2004, Washington, DC.

Setzer, W. 2005. Dose-Response Modeling for Hazard Characterization of Environmental Contaminants. Presentation at the Sixth Meeting on Models in the Regulatory Decision Process, December 2, 2005, Washington, DC.

Shapiro, S.A., R. Steinzor, and M. Clune. 2006. Ossifying Ossification: Why the Information Quality Act Should Not Provide for Judicial Review. CPR White Paper No. 601. February 2006 [online]. Available: http://www.

progressivereform.org/articles/CPR_IQA_601.pdf [accessed April 27, 2007].
Shoemaker, L. 2004. Modeling and Decision Making Overview. Presentation at the First Meeting on Models in the Regulatory Decision Process, March 18, 2004, Washington, DC.
Singer, B.C., and R.A. Harley. 2000. A fuel-based inventory of motor vehicle exhaust emissions in the Los Angeles area during summer 1997. Atmos. Environ. 34(11):1783-1795.
Slaughter, S. 2004. Center for Regulatory Effectiveness' Perspective on Environmental Modeling. Presentation at the First Meeting on Models in the Regulatory Decision Process, March 18, 2004, Washington, DC.
Starfield, A.M., and A.L. Bleloch. 1991. Building Models for Conservation and Wildlife Management, 2nd Ed. Edina, MN: Burgess Press.
Stavins, R.N. 1998. What can we learn from the Grand Policy Experiment? Lessons from SO2 allowance trading. J. Econ. Perspect. 12(3):69-88.
Stedman, D.H. 1989. Automobile carbon monoxide emission. Environ. Sci. Technol. 23(2):147-148.
Stedman, D.H., G.A. Bishop, P. Aldrete, and R.S. Slott. 1997. On-road evaluation of an automobile emission test program. Environ. Sci. Technol. 31(3):927-931.
Swayne, D.F., D. Cook, and A. Buja. 2002. XGobi-A System for Multivariate Data Visualization. AT&T [online]. Available: http://www.research.att.com/areas/stat/xgobi/ [accessed Nov. 27, 2006].
Taylor, G.I. 1915. Eddy motion in the atmosphere. Philos. T. Roy. Soc. A 215:1-26.
Tang, T., M. Roberts, and C. Ho. 2003a. Sensitivity Analysis of MOBILE6 Mobile Source Emissions Factor Model. U.S. Department of Transportation, Federal Highway Administration.
Tang, T., M. Claggett, J. Byun, M. Roberts J. Granell, and D. Aspy. 2003b. MOBILE6.2 Air Toxic Trend and Sensitivity Analysis, U.S. Department of Transportation, Federal Highway Administration.
TCEQ (Texas Commission on Environmental Quality). 2004. Required control strategy elements. Chapter 5 in Revision to the State Implementation Plan (SIP) for the Control of Ozone Air Pollution: Houston/Galveston/Brazoria Ozone Nonattainment Area. Project No. 2004-042-SIP-NR. Texas Commission on Environmental Quality, Austin, TX [online]. Available: http://www.tnrcc.state.tx.us/oprd/sips/june2004hgb_EDrec.html#docs [accessed June 8, 2005].
Thompson, M. 1989. Postscript: Engineering and anthropology: Is there a difference? Pp. 138-150 in Environmental Threats: Perception, Analysis and Management, J. Brown, ed. New York: Belhaven.
Thompson, M., M. Warburton, and T. Hatley. 1986. Uncertainty on a Himalayan Scale. London: Milton Ash Editions/Ethnographica.

Tierney, G. 2004. Decision Making with Mobile Source Models. Presentation at the First Meeting on Models in the Regulatory Decision Process, March 18, 2004, Washington, DC.

Tinbergen, J. 1937. An Econometric Approach to Business Cycle Problems. Paris: Hermann & Co.

Tunkel, J., K. Mayo, C. Austin, A. Hickerson, and P. Howard. 2005. Practical considerations on the use of predictive models for regulatory purposes. Environ. Sci. Technol. 39(7):2188-2199.

van der Sluijs, J.P., J.S. Risbey, P. Kloprogge, J.R. Ravetz, S.O. Funtowicz, S.C. Quintana, Â. Guimarães Pereira, B. De Marchi, A.C. Petersen, P.H.M. Janssen, R. Hoppe, and S.W.F. Huijs. 2003. RIVM/MNP Guidance for Uncertainty Assessment and Communication: Detailed Guidance. Utrecht University, The Netherlands [online]. Available: http://www.nusap.net/downloads/detailedguidance.pdf [accessed April 27, 2007].

van der Sluijs, J.P., P.H.M. Janssen, A.C. Petersen, P. Kloprogge, J.S. Risbey, W. Tuinstra, and J.R. Ravetz. 2004. RIVM/MNP Guidance for Uncertainty Assessment and Communication: Tool Catalogue for Uncertainty Assessment. Report No. NWS-E-2004-37. Utrecht University, The Netherlands [online]. Available: http://www.nusap.net/downloads/toolcatalogue.pdf [accessed April 27, 2007].

Volinsky, C.T., D. Madigan, A.E. Raftery, and R.A. Kronmal. 1997. Bayesian model averaging in proportional hazard models: Assessing the risk of the stroke. Appl. Statist. 46(4):433-448.

Vollenweider, R.A. 1968. Scientific Fundamentals of the Eutrophication of Lakes and Flowing Waters with Particular Reference to Nitrogen and Phosphorus as Factors in Eutrophication. Report DAS/CSI/68.27. Paris: Organization for Economic Cooperation and Development.

Voltaggio, T. 2004. Environmental Modeling – A Regional Perspective. Presentation at the First Meeting on Models in the Regulatory Decision Process, March 18, 2004, Washington, DC.

Wallsten, T.S., D.V. Budescu, A. Rapoport, R. Zwick, and B. Forsyth. 1986. Measuring the vague meaning of probability terms. J. Exp. Psychol. Gen. 115(4):348-365.

Wania, F., and D. Mackay. 1995. A global distribution model for persistent organic chemicals. Sci. Total Environ. 160/161:211-232.

Wania, F., and D. Mackay. 1999. Global chemical fate of ά-hexachlorocyclohexane. 2. Use of a global distribution model for mass balancing, source apportionment, and trend prediction. Environ. Toxicol. Chem. 18(7):1400-1407.

Watson, J.G, J.C. Chow, and E.M. Fujita. 2001. Review of volatile organic compound source apportionment by chemical mass balance. Atmos. Environ. 35(9):1567-1584.

Weart, S. 2003. The Discovery of Global Warming. Cambridge, MA: Harvard University Press.

Weaver, J. 2004. Modeling Leaking Underground Storage Tanks. Presentation at the First Meeting on Models in the Regulatory Decision Process, March 18, 2004, Washington, DC.

Weed, D.L. 2005. Weight of evidence: A review of concept and methods. Risk Anal. 25(6):1545-1557.

Wenzel, T. 2001. Evaluating the long-term effectiveness of the Phoenix IM240 program. Environ. Sci. Policy 4(6):377-389.

Werick, W.J. 1995. National Study of Water Management During Drought. IWR Report 94-NDS-12. Water Resources Center, U.S. Army Corps of Engineers, Fort Belvoir, VA. September 1995.

Woodward, S.M. 1920. Hydraulics of the Miami Flood Control Project. Technical Report Part 7. Dayton, OH: Miami Conservancy District.

Wool, T. 2004. Overview of the TMDL Program and Modeling Approaches. Presentation at the First Meeting on Models in the Regulatory Decision Process, March 18, 2004, Washington, DC.

Zhang, Y., B. Pun, K. Vijayaraghavan, S.Y. Wu, C. Seigneur, S.N. Pandis, M.Z. Jacobson, A. Nenes, and J.H. Seinfeld. 2004. Development and application of the Model of Aerosol Dynamics, Reaction, Ionization and Dissolution. J. Geophys. Res. 109:D01202. doi:10.1029/2003 JD003501.

Glossary

Accuracy – Closeness of a measured or computed value to its "true" value, where the true value is obtained with perfect information. Due to the natural heterogeneity and stochasticity of many environmental systems, this true value exists as a distribution rather than a discrete value. In these cases, the true value will be a function of spatial and temporal aggregation.

Acid Deposition – A comprehensive term for the various ways acidic compounds precipitate from the atmosphere and deposit onto surfaces. It can include (1) wet deposition by means of acid rain, fog, and snow; and (2) dry deposition of acidic particles (aerosols).

Acute Exposure – One or a series of short-term exposures generally lasting less than 24 hours.

Acute Health Effect – A health effect that occurs over a relatively short period of time (e.g., minutes or hours). The term is used to describe brief exposures and effects that appear promptly after exposure.

Air Toxics – Also known as toxic air pollutants or hazardous air pollutants (HAPs) are those pollutants known to or suspected of causing cancer or other serious health problems. The Clean Air Act Amendments of 1990 listed 189 of these air toxics as HAPs because of their potential to be carcinogens, respiratory toxicants, neurotoxicants, or cause other harmful effects. They are differentiated from criteria air pollutants under the air quality management system laid out by the Clean Air Act.

Algorithm – A set of mathematical steps or procedures used for solving a problem.

Ambient Air – The air outside of structures. Often used interchangeably with "outdoor air."

Analytical Models – Models that can be solved mathematically in closed form. For example, some model algorithms that are based on relatively simple differential equations can be solved analytically to provide a single solution.

Application Niche – The set of conditions under which the use of a model is scientifically defensible.

Bayesian analysis – An approach to statistical analysis that is based on Bayes's Theorem, which states that the posterior probability of a parameter *p* is proportional to the prior probability of parameter *p* multiplied by the likelihood of *p* derived from the data collected. The Bayesian approach attempts to keep track of how a priori expectations about some phenomenon of interest can be refined and how observed data can be integrated with such a priori beliefs, to arrive at updated posterior expectations about the phenomenon. The Bayesian approach to decision making incorporates new information or data into the decision process. It allows the analyst to use both sample (data) and prior (expert-judgment) information in a logically consistent manner in making inferences. As further information becomes available, the original assumptions are refined and corrected.

Bias – *Systematic deviation* between a measured (observed) or computed value and its "true" value. Bias is affected by faulty instrument calibration and other measurement errors, systematic errors during data collection, and sampling errors, such as incomplete spatial randomization during the design of sampling programs.

Biologically Based Dose-Response (BBDR) Model – A predictive model that describes biological processes at the cellular and molecular level linking the target organ dose to the adverse effect. BBDR models predict dose-response relationships on the basis of principles of biology, pharmacokinetics, and toxicology.

Boundaries – The spatial and temporal conditions and practical constraints under which environmental data are collected. Boundaries specify the area or volume (spatial boundary) and the time period (temporal boundary) to which a decision will apply.

Boundary Conditions – The physical conditions at the boundaries of a system or at the edges of the region being modeled.

Calibration – The process of adjusting model parameters within physically defensible ranges until the resulting predictions give the best possible fit to the observed data.

Catalytic Converter – A mobile-source emissions-control device designed to reduce emissions of nitrogen oxides, hydrocarbons, and carbon monoxide.

Chronic Exposure – Long-term exposure usually lasting 1 year to a lifetime.

Chronic Health Effect – A health effect that occurs over a relatively long period of time (e.g., months or years).

Clean Air Act (CAA) – Federal legislation administered by EPA that serves as the primary means of regulating ambient air quality in the United States. The original Clean Air Act in the United States was passed in 1963, but most of the national air pollution control program is based on the 1970 version of the law. The 1990 Clean Air Act Amendments are the most recent revisions of the law.

Clean Water Act (CWA) – Federal legislation administered by EPA that serves as the primary means of regulating the surface water quality of the United States. The original legislation was passed in 1972 as the Federal Water Pollution Control Act and became known as the Clean Water Act after Congress passed amendments to it in 1977.

Code – Instructions, written in the syntax of a computer language, which provide the computer with a logical process. Code may also be referred to as "computer program." The term "code" describes the fact that computer languages use a different vocabulary and syntax than algorithms that may be written in standard language.

Code of Federal Regulations (CFR) – Document that codifies all rules of the executive departments and agencies of the federal government. It is divided into 50 volumes, known as titles. Title 40 of the CFR (referenced as 40 CFR) lists all environmental regulations.

Community Multiscale Air Quality (CMAQ) Model – An air quality model designed to simulate and model a wide range of physical and chemical processes relating to air quality at particular scales in the lower atmosphere over a regional and subregional scale.

Computational Model – A model that is expressed in formal mathematics using equations, statistical relationships, or a combination of the two. Although values, judgment, and tacit knowledge are inevitably embedded in the structure, assumptions, and default parameters, computational models are inherently quantitative, relating phenomena through mathematical relationships and producing numerical results.

Computational Toxicology – The application of mathematical and computer models to predict the effect of an environmental agent and elucidate the cascade of events that result in an adverse response. It uses technologies developed in computational chemistry (computer-assisted simulation of molecular systems), molecular biology (characterization of genetics, protein synthesis, and molecular events involved in biological response to an agent), bioinformatics (computer-assisted collection, organization, and analysis of large data sets of biological information), and systems biology (mathematical modeling of biological systems and phenomena). The goals of using computational toxicology are to set priorities among chemicals on the basis of screening and testing data and to develop predictive models for quantitative risk assessment.

Conceptual Model – An abstract representation that provides the general structure of a system and the relationships within the system that are known or hypothesized to be important. Many conceptual models have as a key component a graphical or pictorial representation of the system.

Contaminant – A substance that is either present in an environment where it does not belong or is present at levels that might cause harmful (adverse) health effects.

Corroboration (Model) – Quantitative and qualitative methods for evaluating the degree to which a model corresponds to reality. In some disciplines, this process has been referred to as validation. In general, the term "corroboration" is preferred because it implies a claim of usefulness and not truth.

Criteria Air Pollutants – An air pollutant for which National Ambient Air Quality Standards have been set. There are six common air pollutants (carbon monoxide, lead, nitrogen dioxide, ozone, particulate matter, and sulfur dioxide) that have been designated as criteria pollutants. The Clean Air Act states that the presence of criteria pollutants in the ambient air results from numerous or diverse mobile or stationary sources.

Cumulative Risk – The combined risks from aggregate exposures to multiple agents or stressors.

Design Standard – A technology-based standard that requires emitters to use a specific technology to control emissions of a pollutant. These can also be called engineering standards.

Deterministic Model – A mathematical model that contains no random (stochastic) components; consequently, each component and input is determined exactly. Because this type of model does not explicitly simulate the effects of data uncertainty or variability, changes in model outputs are solely due to changes in model components.

Domain (Spatial and Temporal) – The limits of space and time that are specified within a model's boundary conditions (see Boundary Conditions).

Domain Boundaries (Spatial and Temporal) – The spatial and temporal domain of a model are the limits of extent and resolution with respect to time and space for which the model has been developed and over which it should be evaluated.

Dose – The amount of a contaminant that is absorbed or deposited in the body of an exposed person for an interval of time—usually from a single medium. Total dose is the sum of doses received by interactions with all environmental media that contain the contaminant. Units (mass) of dose and total dose are often converted to units of mass or contaminant per volume of physiological fluid or mass of tissue.

Dose-Response Relationship – The relationship between a quantified exposure (or dose) and a quantified effect

Emission Rate – The weight of a pollutant emitted per unit of time (e.g., tons/year).

Emissions Budget – Allowable emissions levels identified as part of a state implementation plan for pollutants emitted from mobile, industrial, stationary, and area sources. These emissions levels are used for meeting emission-reduction milestones, attainment, or maintenance demonstrations.

Emissions Factor – For mobile sources, the emission factor is the relationship between the amount of pollution produced and the number of vehicle miles traveled. For stationary sources, the relationship between the amount of pollution produced and the amount of raw material processed or burned. By using the emission factor of a pollutant and specific data regarding activities (quantities of materials used by a given source or number of miles traveled), it is possible to compute emissions for the source.

Emissions Inventory – An estimate of the amount of a pollutant emitted into the atmosphere from major mobile, stationary, areawide, and natural sources over a specific period of time, such as a day or a year.

Empirical Model – An empirical model is one where the structure is determined by the observed relationship among experimental data. These models can be used to develop relationships that are useful for forecasting and describing trends in behavior but may not necessarily be mechanistically relevant.

Environmental Regulatory Model – A computational model used to inform the environmental regulatory process. Some models are independent of a specific regulation, such as water quality or air quality models that are used in an array of application settings. Other models are created to provide a regulation-specific set of analyses completed during the development and assessment of specific regulatory proposals. The approaches can range from single parameter linear relationship models to models with thousands of separate components and many billions of calculations.

Epidemiology – The study of the distribution and determinants of disease or health status in a population; the study of the occurrence and causes of health effects in humans.

Evaluation (Model) – The process used to generate information to determine whether a model and its results are of a quality sufficient to serve as the basis for a regulatory decision.

Evaporative Emissions – Hydrocarbon emissions that do not come from the tailpipe of a car but come from evaporation, permeation, seepage, and leaks in a vehicle's fueling system. The term is sometimes used interchangeably with *nontailpipe emissions*.

Ex Ante – Analysis of the effects of a policy based only on information available before the policy is undertaken. Also termed prospective analysis.

Ex Post – Analysis of the effects of a policy based on information available after the policy has been implemented and its performance observed. Also termed retrospective analysis.

Exceedance – An air pollution event in which the ambient concentration of a pollutant exceeds the National Ambient Air Quality Standards.

Expert Elicitation – A process for obtaining expert beliefs about subjective quantities and probabilities. Typically, structured interviews and questionnaires are used to elicit the necessary knowledge. Expert elicitations may also include "coaching" techniques to help the expert conceptualize, visualize, and quantify the knowledge being sought.

Exposure – Contact between an agent and a target. Contact takes place at an exposure surface over an exposure period, which is the time of continuous contact between an agent and a target

Exposure Assessment – The process of characterizing the magnitude, frequency, and duration of exposure to an agent, along with the number and characteristics of the population exposed. Ideally, it describes the sources, pathways, routes, and uncertainties in the assessment.

Exposure Pathway – The course a substance takes from its source (where it began) to its end point (where it ends), and how people can come into contact with (or get exposed to) it. An exposure pathway has five parts: a source of contamination (such as an abandoned business); an environmental media and transport mechanism (such as movement through groundwater); a point of exposure (such as a private well); a route of exposure (eating, drinking, breathing, or touching), and a receptor population (people potentially or actually exposed). When all five parts are present, the exposure pathway is termed a completed exposure pathway.

Genomics – The study of genes and their function.

Greenhouse Gas – Atmospheric gases such as carbon dioxide, methane, chlorofluorocarbons, nitrous oxide, ozone, and water vapor that slow the passage of re-radiated heat through the earth's atmosphere.

Hazard Assessment – The process of determining whether exposure to an agent can cause an increase in the incidence or severity of a particular health effect (e.g., cancer, birth defect).

Hazardous Air Pollutants (HAPs) – Air toxics listed under section 112(b) of the Clean Air Act Amendments of 1990.

Informatics (Bioinformatics) – The science of managing and analyzing vast amounts of biological data using advanced computing techniques. Especially important in analyzing genomic research data.

Marginal Benefit – The additional benefit gained from one more unit of output. In terms of reducing emissions, it represents the benefits from reducing emissions by one more unit.

Marginal Cost – The additional cost associated with producing one more unit of output. In terms of reducing emissions, it represents the cost of reducing emissions by one more unit.

Model – A simplification of reality that is constructed to gain insights into select attributes of a particular physical, biological, economic, or social system. Models can be of many different forms. They can be computational. Computational models include those that express the relationships among components of a system using mathematical relationships. They can be physical, such as models built to analyze effects of hydrodynamic or aeronautical conditions or to represent landscape topography. They can be empirical, such as statistical models used to relate chemical properties to molecular structures or human dose to health responses. Models also can be analogs, such as when nonhuman species are used to estimate health effects on humans. And they can be conceptual, such as a flow diagram of a natural system showing relationships and flows among individual components in the environment or a business model that broadly shows the operations and organization of a business. The above definitions are not mutually exclusive. For example, a computational model may be developed from conceptual and physical models, and an animal analog model can be the basis for an empirical model of human health impacts.

Module – An independent or self-contained component of a model that is used in combination with other components and forms part of one or more larger programs.

National Ambient Air Quality Standards (NAAQS) – Standards set by EPA for the maximum levels of criteria air pollutants that can exist in the outdoor air without adverse effects on human health or the public welfare. There are four elements of a NAAQS (1) the pollutant indicator (such as $PM_{2.5}$), (2) the concentration of the indicator in the air, (3) the time over which measurements are made or averaged, and (4) the statistical form of the standard used to determine the allowable number of exceedances (such as the fourth highest value over a 3-year period).

National Pollution Discharge Elimination System (NPDES) – Federal regulations that regulate discharge of wastewater to surface waters, such as streams, rivers, lakes, and estuaries. An NPDES permit is required for any project involving the construction, alteration, and/or operation of any sewer system, treatment works, or disposal system and for construction of certain storm water runoff structures that would result in a discharge into surface waters.

Nonattainment Area – A geographic area designated by EPA to have concentrations of a criteria pollutant in excess of the NAAQS. A single geographic area may have acceptable levels of some criteria air pollutants but unacceptable levels of others; thus, an area can be both an attainment area for one pollutant and a nonattainment area for another.

Nonpoint Source Pollution – Sources of water pollution not associated with a distinct discharge source; includes rainwater, erosion, run-off from roads, farms, and parking lots, and seepage from soil-based wastewater disposal systems.

Parameters – Terms in the model that are fixed during a model run or simulation but can be changed in different runs as a method for conducting sensitivity analysis or to achieve calibration goals.

Photochemical Reaction – A term referring to a chemical reaction brought about by sunlight, such as the formation of ozone from the interaction of oxygen and nitrogen oxides and/or hydrocarbons in the presence of sunlight.

Physiologically Based Pharmacokinetic (PBPK) Model – A model that estimates the dose to a target tissue or organ by taking into account the rate of absorption into the body, distribution among target organs and tissues, metabolism, and excretion.

Plume – A volume of a substance that moves from its source to places farther away from the source. Plumes can be described by the volume of air or water they occupy and the direction they move. For example, a plume can be a column of smoke from a chimney or a substance moving with groundwater.

Point Source Pollution – A specific discharge to a water body, ambient air, or land that is traceable to a distinct source (e.g., pipe, smokestack, and container) such as those from wastewater treatment plants, power plants, or industrial facilities.

Precision – The quality of being reproducible in amount or performance. With models and other forms of quantitative information, precision refers specifically to the number of decimal places to which a number is computed as a measure of the preciseness or exactness with which a number is computed.

Proteomics – The study of the full set of proteins encoded by a genome.

Regulatory Impact Analysis (RIA) – An analysis document produced by EPA for each major rulemaking listing the expected impacts of the rule, including environmental impacts, health impacts, cost–benefit analyses, economic impacts, and small business impacts.

Reliability – The confidence that (potential) users have in a model and in the information derived from the model such that they are willing to use the model and the derived information. Specifically, reliability is a function of the performance record of a model and its conformance to best available, practicable science.

Risk Assessment (in the context of human health) – The evaluation of scientific information on the hazardous properties of environmental agents (hazard identification), the dose-response relationship (dose-response assessment), and the extent of human exposure to those agents (exposure assessment). The product of the risk assessment is a statement describing the populations or individuals that are likely to be harmed and to what degree (risk characterization).

Risk Characterization (in the context of human health) – The integration of information on hazard, dose-response, and exposure to provide an estimate of the likelihood that any of the identified adverse effects will occur in exposed people.

Risk Management (in the context of human health) – A decision-making process that accounts for political, social, economic, and engineering implications together with risk-related information to develop, analyze, and compare management options and select the appropriate managerial response to a potential adverse health risk.

Robustness – The capacity of a model to perform equally well across the full range of environmental conditions for which it was designed.

Safe Drinking Water Act (SDWA) – Legislation to ensure safe drinking water. Passed by Congress in 1974 and amended in 1986, it directs EPA to establish and enforce water quality standards to protect public health.

Screening Model – A type of model designed to provide a "conservative" or risk-averse answer. Because screening models can be used with limited information and are conservative, they can be used to determine whether more refined models would be useful or whether the screening model results are sufficient to make decisions without proceeding to a refined model.

Sensitivity – The degree to which the model outputs are affected by changes in a selected input parameters.

State Implementation Plan (SIP) – A detailed description of the scientific methods and emission-reduction programs a state will use to carry out its responsibilities under the Clean Air Act for complying with the NAAQS. The Clean Air Act requires that EPA approve each SIP after the public has had an opportunity to participate in its review and approval.

Stochastic Model – A model that includes variability (see definition) in model parameters. This variability is a function of (1) changing environmental conditions, (2) spatial and temporal aggregation within the model framework, and (3) random variability. The solutions obtained by the model or output is therefore a function of model components and random variability.

Susceptibility – Increased likelihood of an adverse effect, often discussed in terms of relationship to a factor that can be used to describe a human subpopulation (e.g., life stage, demographic feature, and genetic characteristic).

Susceptible Subgroups – May refer to life stages (e.g., children and the elderly) or to other segments of the population (e.g., people who have asthma or who are immune compromised), but are likely to be chemical-specific and may not be consistently defined in all cases.

Technology-Based Standards – A type of standard that dictates polluters use specific techniques (e.g., a particular type of pollution abatement equipment) or follow a specific set of operating procedures and practices.

Technology Forcing – The establishment by a regulatory agency of a requirement to achieve an emissions limit, within a specified time frame, that can be reached through use of unspecified technology or technologies that have not yet been developed for widespread commercial applications and have been shown to be feasible on an experimental or pilot-demonstration basis.

Total Maximum Daily Load (TMDL) – The total waste (pollutant) loading from point and nonpoint sources that a water body can assimilate while still maintaining its water quality classification and standards.

Toxic Substances Control Act (TSCA) – Federal legislation administered by EPA that regulates the manufacture, labeling, and distribution of chemicals outside of pesticides and drugs. It requires tests of chemicals that may harm human health or the environment, reviews of new chemical substances, limits on the availability of some existing chemicals, and import certification standards to ensure that imported chemicals comply with domestic rules.

Toxicogenomics – The study of how genomes respond to environmental stressors or toxicants. Combines genomewide mRNA expression profiling with protein expression patterns using bioinformatics to understand the role of gene-environment interactions in disease and dysfunction.

Toxicology – The study of the harmful effects of substances on living organisms.

Transparency – The clarity and completeness with which data, assumptions and methods of analysis are documented.

Variability – Observed differences attributable to *true heterogeneity* or diversity and the result of natural random processes—usually not reducible by further measurement or study (although it can be better characterized).

Water Quality Criteria – Levels of water quality expected to render a water body suitable for its designated use. Criteria are based on specific levels of pollutants that would make the water harmful if used for drinking, swimming, fish production, or industrial uses.

Water Quality Standards – Ambient standards for water bodies adopted by a state and approved by EPA that prescribe the use of the water body and establish the water quality criteria that must be met to protect designated uses. Water quality standards may apply to dissolved oxygen, heavy metals, pH, and other water constituents.

SOURCES

CARB at http://www.arb.ca.gov/html/gloss.htm.
EPA at http://www.epa.gov/oar/oaqps/peg_caa/pegcaa10.html.
at http://www.epa.gov/oms/stds–ld.htm.
at http://www.epa.gov/ord/crem/library/CREM%20Modeling%20Glossary%2012_03.pdf.
at http://www.epa.gov/iris/gloss8.htm.
Human Genome Project at http://www.ornl.gov/sci/techresources/Human_Genome/glossary/glossary.shtml; Agency for Toxic Substances and Disease Registry at http://www.atsdr.cdc.gov/glossary.html.

Appendix A

Biographical Information on the Committee on Models in the Regulatory Decision Process

Chris G. Whipple is a principal in ENVIRON International Corporation in Emeryville, California, which provides consulting services mainly to private industry. His professional interests are in risk assessment, and he often works on environmental and health issues involving radioactive materials or mercury. He is currently a member of the National Research Council (NRC) Board on Environmental Studies and Toxicology (BEST) and chairs the BEST work group on Environmental Sciences and Engineering. He also currently chairs the Committee on Medical Isotope Production Without Highly Enriched Uranium. He previously served as chair of the Board on Radioactive Waste Management (BRWM). He has served on and chaired numerous other NRC committees. Dr. Whipple received his B.S. in engineering science from Purdue University and his M.S. and Ph.D. in engineering science from the California Institute of Technology.

M. Bruce Beck is Wheatley-Georgia Research Alliance professor of environmental systems analysis at the University of Georgia. His research interests include modeling and simulation of surface-water quality; control and automation of wastewater treatment processes; and novel systems for urban wastewater infrastructures. Before his appointment at the University of Georgia, he was a visiting research fellow at the Lund Institute of Technology, Sweden; a Royal Society Ernest Cook Trust research fellow at Cambridge University in England; a research scholar at the International Institute for Applied Systems Analysis in Austria; and a member of faculty at Imperial College, London, where he is visiting professor. Dr. Beck received his B.Sc. in chemical engineering from the University of Exeter and his Ph.D. in control engineering from the University of Cambridge.

Clayton J. Clark II is an assistant professor of civil and coastal engineering at the University of Florida. His research interests include implementing waste management procedures; delineating and remediation of contaminated hazardous waste sites; and combining chemical and environmental engineering techniques for hazardous waste handling, disposal, and treatment from both soil and aqueous systems. Dr. Clark is also associated with the Florida Center for Solid and Hazardous Waste Management through his work on the modeling of the fate and transport of contaminants from pressure-treated wood. Dr. Clark received his B.S. from Florida A&M University and his Ph.D. from the University of Florida.

Robert T. Clemen is an associate professor of decision sciences at the Fuqua School of Business, Duke University. Before going to Duke, Dr. Clemen was associate professor at the University of Oregon and senior researcher at Decision Research in Eugene, Oregon. His teaching and research interests focus on decision analysis, especially the use of models and expert judgment for decision making. He received his B.A. from Stanford University, M.B.A. from the University of Colorado, and Ph.D. from Indiana University.

Judith A. Graham is the managing director of the American Chemistry Council's (ACC) long-range research initiative (LRI). The LRI sponsors research to advance the science of risk assessment for the health impacts of chemicals to support decision making by government, industry, and the public. Her research interests include inhalation toxicology, exposure analysis, and health effects and health risks of air pollutants. Before joining ACC, Dr. Graham was associate director for health at EPA's National Exposure Research Laboratory (NERL). She has served on several previous National Research Council study committees. Dr. Graham received her Ph.D. in physiology and pharmacology from Duke University.

Louis J. Gross is the director of the Institute for Environmental Modeling, a professor of ecology and evolutionary biology, and a professor of mathematics at the University of Tennessee in Knoxville. His research interests include mathematical ecology, computational ecology, quantitative training for life science students, photosynthetic dynamics, and parallel computation for ecological models. He is currently president of the Society for Mathematical Biology and recently chaired the NRC Committee on Integrating Education and Biocomplexity Research. He received his B.S. degree from Drexel University and his Ph.D. in Applied Mathematics from Cornell University.

Winston Harrington is a senior fellow at Resources for the Future (RFF). His research interests include urban transportation, motor vehicles and air quality, and estimating the costs of environmental policies. He has worked on the economics of enforcing environmental regulations, the health benefits of improved air quality, and the costs of waterborne disease outbreaks. He received his A.B. in mathematics from the University of North Carolina at Chapel Hill, his M.A. in mathematics from Cornell University, and his Ph.D. in city and regional planning from the University of North Carolina at Chapel Hill.

Philip Howard is a senior director at Syracuse Research Corporation. His expertise is in exposure and risk assessment, environmental fate and transport modeling, and the evaluation of data related to the physical and chemical properties of chemicals. He directed the design and maintenance of Syracuse Research Corporation's Environmental Fate Database. He also directed the information evaluation and peer review of the Environmental Fate and Exposure section of the National Library of Medicine's Hazardous Substance Data Bank. In addition, he is the director of a project for EPA to review data registration packages on the chemistry and fate of pesticides in the environment to identify variances from published guidelines, standard evaluation practices, and data review guidelines.

Kimberly L. Jones is associate professor of civil engineering at Howard University. Her research interests include physical-chemical treatment processes, membrane processes, adsorption, mass transport, interfacial phenomenon, water and wastewater treatment plant design, and water quality. Dr. Jones also is the deputy director of the Keck Center for the Design of Nanoscale Materials for Molecular Recognition. She received her B.S. in civil engineering from Howard University, her M.S. in civil engineering from the University of Illinois, and her Ph.D. in environmental engineering from Johns Hopkins University.

Thomas E. McKone is a senior scientist and deputy department head at the Ernest Orlando Lawrence Berkeley National Laboratory and is an adjunct professor in the School of Public Health at the University of California, Berkeley. His research interests include the chemical transport and accumulation of toxic chemicals in multiple environmental media (air, water, soil), developing multimedia compartment models that can be used in quantitative risk assessments, and human exposure and health risk assessment. Dr. McKone has served on previous NRC study committees. He received his M.S. and Ph.D. in engineering from the University of California at Los Angeles.

Naomi Oreskes is an associate professor in the Department of History at the University of California, San Diego, where she also directs the Program in Science Studies. Her research focuses on the historical development of scientific knowledge, methods, and practices in the earth and environmental sciences. Dr. Oreskes has also been a visiting associate professor in the Department of History of Science at Harvard University. Before going to the University of California, she was an associate professor of history and philosophy of science at New York University. She was a member of the NRC Committee for the International Union of Geological Sciences. She received her B.Sc. from University of London in England and her Ph.D. from Stanford University.

Spyros N. Pandis is professor of chemical engineering at the University of Patras, Greece, and Elias Research Professor of chemical engineering and engineering and public policy at Carnegie Mellon University. His research interests include atmospheric chemistry, atmospheric pollution modeling, aerosol science, global change, and environmental policy analysis. He is serving on the NRC Committee on Air Quality Management in the United States and is a former member of the committee reviewing the U.S. Department of Energy Office of Fossil Energy's research plan for fine particulates. Dr. Pandis received his Ph.D. in chemical engineering from the California Institute of Technology.

Louise M. Ryan is chair of the Biostatistics Department at Harvard School of Public Health. Her research is on statistical methods related to environmental health research and risk assessment. She has served on advisory boards for several government agencies, including the National Toxicology Program and EPA. Dr. Ryan has served on several previous NRC study committees. She received her Ph.D. from Harvard University.

Michael L. Stein is a professor of statistics at the University of Chicago and director of the Center for Integrating Statistical and Environmental Science. His research interests focus on statistical models and methods for spatial and spatial-temporal processes. He is interested in the nature of the spatial-temporal interactions implied by these models and on developing statistical methods for assessing these interactions. Dr. Stein received his B.S. in mathematics from the Massachusetts Institute of Technology and his M.S. and Ph.D. in statistics from Stanford University.

Wendy E. Wagner is a professor of law at the University of Texas at Austin. Before entering teaching, she practiced for 4 years, first as an honors attorney in the Enforcement Division of the U.S. Department of Justice's Environment and Natural Resources Division and then as pollution control coordinator in the U.S. Department of Agriculture Office of the General Counsel. Professor Wagner received her M.S. in environmental studies and her law degree from Yale University. She currently serves as an officer in the Society for Risk Analysis and the American Bar Association's Administrative Law Section and is a Member Scholar of the Center for Progressive Regulation.

Appendix B

Public Workshop Presentations to the Committee on Models in the Regulatory Decision Process

March 18, 2004, National Academy of Sciences Main Building, Washington, DC

U.S. EPA's Council for Regulatory Environmental Modeling
Gary Foley, Office of Research and Development, Council on Regulatory Environmental Modeling, U.S. Environmental Protection Agency

Environmental Modeling—A Regional Perspective
Tom Voltaggio, Deputy Administrator, Region 3, U.S. Environmental Protection Agency

Environmental Economic Models
Albert McGartland, National Center for Environmental Economics, U.S. Environmental Protection Agency

Using Air Quality Models for Emissions Management Decisions—Making Decisions in the Face of Uncertainty
S.T. Rao, Atmospheric Sciences Modeling Division, U.S. Environmental Protection Agency

Model Use in the Office of Prevention, Pesticides and Toxic Substances
Joseph Merenda, Office of Prevention, Pesticides, and Toxic Substances, U.S. Environmental Protection Agency

Modeling Leaking Underground Storage Tanks
Jim Weaver, National Exposure Research Laboratory, U.S. Environmental Protection Agency

Presentation to the Committee on Environmental Decision Making Principles and Criteria for Models
David Burden, Groundwater Technical Support Center, U.S. Environmental Protection Agency

Overview of the TMDL Program & Modeling Approaches
Tim Wool, National Exposure Research Laboratory, Region 4, U.S. Environmental Protection Agency

Modeling and Decision Making Overview
Leslie Shoemaker, Tetra Tech, Inc.

State Perspectives on Modeling in Support of TMDL Development
Jim George, Maryland Department of the Environment

Estimating Motor Vehicle Emissions: A Tale of 2 Models
Cecilia Ho, Federal Highway Administration

Corps of Engineers Planning Models Improvement Program
Harry Kitch, Planning & Policy Division, U.S. Environmental Protection Agency

Models in the Regulatory Decision Process
Timothy Miller, National Water Quality Assessment Program, U.S. Geological Survey

Comments on Behalf of the Natural Resources Defense Council
Jennifer Sass, Natural Resources Defense Council

Comments on Behalf of the Center for Regulatory Effectiveness
Scott Slaughter, Center for Regulatory Effectiveness

Guidelines for Model Choice in a Regulatory Environment: Sound Science Reflecting Sound Values

Adam Finkel, Senior Safety and Health Adviser, Office of the Assistant Secretary, U.S. Occupational Safety and Health Administration

Decision Making with Mobile Source Models

Gene Tierney, Director, Center for Air Quality and Modeling, U.S. Environmental Protection Agency

December 2, 2004, Keck Center of the National Academies, Washington, DC

Methods and Applications of Uncertainty and Sensitivity Analysis for Models Used in Regulatory Processes

H. Christopher Frey, Professor, Department of Civil, Construction, and Environmental Engineering, North Carolina State University

Office of Management and Budget Proposed Bulletin on Peer Review and Information Quality

Margo Schwab and John Graham, U.S. Office of Management and Budget

Building Model Confidence and Quality Considerations for Regulatory Decision Makers

Rob Howard, Bechtel-SAIC, LLC

Peer Review of Regulatory Models—The Hard Look and the Long View

Sheila Jasanoff, Centre for Population Health Risk Assessment, Harvard University

Uncertainties in Health Risk Projection Models: Implications for Risk Management

Daniel Krewski; Jan M. Zielinski; Tim Ramsay and Richard T. Burnett, McLaughlin Centre for Population Health Risk Assessment

Interagency Steering Committee on Multimedia Environmental Modeling: Uncertainty Analysis Issues

George Leavesley, Chair, Interagency Steering Committee on Multimedia Environmental Models, U.S. Geological Survey

EPA Use of ICF's Integrated Planning Model
Sam Napolitano and Elliot Lieberman, Clean Air Markets Division, U.S. Environmental Protection Agency

Parameter and Model Uncertainty in Models for Regulatory Decision Making: Problems and Opportunities
M. Granger Morgan, Department of Engineering and Public Policy, Carnegie Mellon University

Wresting Regulatory Decisions from an Uncertain World
Pasky Pascual, Council for Regulatory Environmental Modeling, U.S. Environmental Protection Agency

Why Use Proprietary Software for Risk Assessment?
Barbara Petersen, Principal Scientist Practice Director, Exponent, Inc., and Durango Software, LLC

Expectations Regarding Peer Review of Especially Significant Regulatory Information
James D. Schaub, Office of Risk Assessment and Cost-Benefit Analysis, U.S. Department of Agriculture

December 2, 2005, National Academy of Sciences Main Building, Washington, DC

Dose-Response Modeling for Hazard Characterization of Environmental Contaminants
Woodrow Setzer, U.S. Environmental Protection Agency

The Use of Physiologically Based Pharmacokinetic Modeling to Improve Chemical Dosimetry in Environmental Risk Assessments
Harvey Clewell, Centers for Health Research

Computational Systems Biology: The Integration of Data Across Multiple Levels of Biological Organization to Understand How Perturbations of Normal Biology Become Adverse Health Effects
Rory Conolly, Center for Computational Toxicology, U.S. Environmental Protection Agency

Global Earth Observation System of Systems (GEOSS) and Environmental Regulatory Modeling
Gary Foley, U.S. Environmental Protection Agency

Making Regulatory Choices: What Does Uncertainty Have to Do With It
Richard Morgenstern, Resources for the Future

Use of Modeling in Decision-Making at EPA
Robert Perciasepe, Audubon Society

Integrating Modeling and Monitoring in Adaptive Strategies to Attain Water Quality Standards
Kenneth Reckhow, Duke University

Computation in the Regulatory Environment: Solution or the Problem
Paul Gilman, Oak Ridge Center for Advanced Studies

Appendix C

Categories of Environmental Regulatory Models

As discussed in Chapter 2, models can be categorized according to their fit into a continuum of processes that translate human activities and natural systems interactions into human health and environmental impacts (see Figure 2-1). The categories of models that are integral to environmental regulation include human activity models, natural systems models, emissions models, fate and transport models, exposure models, human health and environmental response models, economic impact models, and noneconomic impact models. Examples of models in each of these categories are discussed below.

HUMAN ACTIVITY MODELS

Anthropogenic emissions to the environment are inherently linked to human activities. Activity models simulate the human activities and behaviors that result in pollutants. In the environmental regulatory modeling arena, examples of modeled activities are the following:

- Demographic information, such as the magnitude, distribution, and dynamics of human populations, ranging from national growth projections to local travel activity patterns on the order of hours.
- Economic activity, such as the macroeconomic estimates of national economic production and income, final demands for aggregate

industrial sectors, prices, international trade, interest rates, and financial flows.

- Human consumption of resources, such as gasoline or feed, may be translated into pollutant releases, such as nitrogen oxides or nutrients. Human food consumption is also used to estimate exposure to pollutants such as pesticides. Resource consumption in dollar terms may be used to assess economic impacts.
- Distribution and characteristics of land use are used to assess habitat, impacts on the hydro-geologic cycle and runoff, and biogenic pollutant releases.

Human Activity Models

Model	Type	Use	Additional Information
TRANSCAD, TRANPLAN, MinUTP	Travel demand forecasting models	Develops estimations of motor vehicle miles traveled for use in estimating vehicle emissions. Can be combined with geographic information systems (GIS) for providing spatial and temporal distribution of motor vehicle activity.	Caliper Corporation 2007
DRI	Forecasts national economic indicators	Model can forecast over 1,200 economic concepts including aggregate supply, demand, prices, incomes, international trade, interest rates, etc. The eight sectors of the model are: domestic spending, domestic income, tax sector, prices, financial, international trade, expectations, and aggregate supply.	EIA 1993
E-GAS	National and regional economic activity model	Emissions growth factors for various sector for estimating volatile organic compounds, nitrogen oxides, and carbon monoxide emissions.	Young et al. 1994
YIELD	Crop-growth yield model	Predicts temporal and spatial crop yield.	Hayes et al. 1982

NATURAL SYSTEMS PROCESS AND EMISSIONS MODELS

Natural systems process and emissions models simulate the dynamics of ecosystems that directly or indirectly give rise to fluxes of nutrients and other environmental emissions.

Natural Systems Process and Emissions Models

Model	Type	Use	Additional Information
Marine Biological Laboratory General Ecosystem Model (MBL-GEM)	Plot-scale nutrient cycling of carbon and nitrogen	Simulates plot-level photosynthesis and nitrogen uptake by plants, allocation of carbon and nitrogen to foliage, stems, and fine roots, respiration in these tissues, turnover of biomass through litter fall, and decomposition of litter and soil organic matter.	MBL 2005
BEIS	Natural emissions of volatile organic compounds	Simulates nitric oxide emissions from soils and volatile organic compound emissions from vegetation. Input to grid models for NAAQS attainment (CAA).	EPA 2006a Vukovich and Pierce 2002
Natural Emissions Model	Natural emissions of methane and nitrous oxide	Models methane and nitrous oxide emissions from the terrestrial biosphere to atmosphere.	MIT 2006, Sokolov et al. 2005

EMISSIONS MODELS

These models estimate the rate or the amount of pollutant emissions to water bodies and the atmosphere. The outputs of emission models are used to generate inventories of pollutant releases that can then serve as an input to fate and transport models.

Emissions Models

Model	Type	Use	Additional Information
PLOAD	Releases to water bodies	GIS bulk loading model providing annual pollutant loads to waterbodies. Conducts simplified analyses of sediment issues, including a bank erosion hazard index.	EPA 2007a EPA 2001

(Continued)

Emissions Models Continued

Model	Type	Use	Additional Information
SPARROW	Releases to water bodies	Relates nutrient sources and watershed characteristics to total nitrogen. Predicts contaminant flux, concentration, and yield in streams. Provides empirical estimates (including uncertainties) of the fate of contaminants in streams.	USGS 2007a Schwarz et al. 2006
MOBILE MOVES NONROAD	Releases to air	Factors and activities for anthropogenic emissions from mobile sources. Estimates current and future emissions (hydrocarbons, carbon monoxide, nitrogen oxides, particulate matter, hazardous air pollutants, and carbon dioxide) from highway motor vehicles. Model used to evaluate mobile source control strategies, control strategies for state implementation plans, and for developing environmental impact statements, in addition to other research.	EPA 2007b EPA 2006b EPA 2004, EPA 2005a Glover and Cumberworth 2003

FATE AND TRANSPORT MODELS

Fate and transport models calculate the movement of pollutants in the environment. A large number of EPA models fall into this category. They are further categorized into the transport media they represent: subsurface, air, and surface water. In each medium, there are a range of models with respect to their complexity, where the level of complexity is a function of the following:

- The number of physical and chemical processes considered.
- The mathematical representation of those processes and their numerical solution.
- The spatial and temporal scales over which the processes are modeled.

Even though some fate and transport models can be statistical models, the majority is mechanistic (also referred to as process-based models). Such models simulate individual components in the system and the mathematical relationships among the components. Fate and transport model output has traditionally been deterministic, although recent focus on uncertainty and variability has led to some probabilistic models.

Subsurface Models

Subsurface transport is governed by the heterogeneous nature of the ground, the degree of saturation of the subsurface, as well as the chemical and physical properties of the pollutants of interest. Such models are used to assess the extent of toxic substance spills. They can also assess the fate of contaminants in sediments. The array of subsurface models is tailored to particular application objectives, for example, assessing the fate of contaminants leaking from underground gasoline storage tanks or leaching from landfills. Models are used extensively for site-specific risk assessments; for example, to determine pollutant concentrations in drinking-water sources. The majority of models simulate liquid pollutants; however, some simulate gas transport in the subsurface.

Subsurface Models

Model	Type	Use	Additional Information
MODFLOW	3D finite difference for ground water transport	Risk Assessments (RBCA) Superfund Remediation (CERCLA). Modular three-dimensional model that simulates ground water flow. Model can be used to support groundwater management activities.	USGS 2007b Prudic et al. 2004, Wilson and Naff 2004
PRZM	Hydrogeological	Pesticide leaching into the soil and root zone of plants (FIFRA). Estimates pesticide and nitrogen fate in the crop zone root and can simulate soil temperature, volatilization and vapor phase transport in soil, irrigation, and microbial transformation.	EPA 2007c EPA 2005b
BIOPLUME	Two-dimensional finite difference	Simulates organic contaminants in groundwater due to natural processes of	EPA 2006c EPA 1998

(Continued)

Subsurface Models Continued

Model	Type	Use	Additional Information
	and Method of Characteristics (MOC) model	dispersion, advection, sorption, and biodegradation. Simulates aerobic and anaerobic biodegradation reactions.	

Surface Water Quality Models

Surface water quality models are often related to, or are variations of, hydrological models. The latter are designed to predict flows in water bodies and runoff from precipitation, both of which govern the transport of aqueous contaminants. Of particular interest in some water quality models is the mixing of contaminants as a function of time and space, for example, following a point-source discharge into a river. Other features of these models are the biological, chemical, and physical removal mechanisms of contaminants, such as degradation, oxidation, and deposition, as well as the distribution of the contaminants between the aqueous phase and organisms.

Surface Water Quality Models

Model	Type	Use	Additional Information
HSPF	Combined watershed hydrology and water quality	Total maximum daily load determinations TMDL (CWA). Watershed model simulating nonpoint pollutant load and runoff, fate and transport processes in streams.	EPA 2006d Donigian 2002
WASP	Compartment modeling for aquatic systems	Supports management decisions by predicting water quality responses to pollutants in aquatic systems. Multicompartment model that examines both the water column and underlying benthos.	EPA 2006e Brown 1986 Brown and Barnwell 1987
QUAL2E	Steady-state and quasi-dynamic water quality model	Stream water quality model used as a planning tool for developing TMDLs. The model can simulate nutrient cycles, benthic and carbonaceous demand, algal production, among other parameters.	Birgand 2004 Brown 1986, Brown and Barnwell 1987

Air Quality Models

The fate of gaseous and solid particle pollutants in the atmosphere is a function of meteorology, temperature, relative humidity, other pollutants, and sunlight intensity, among other things. Models that simulate concentrations in air have one of three general designs: plume models, grid models, and receptor models. Plume models are used widely for permitting under requirements to assess the impacts of large new or modified emissions sources on air quality or to assess air toxics (HAPs) concentrations close to sources. Plume models focus on atmosphere dynamics. Grid models are used primarily to assess concentrations of secondary criteria pollutants (e.g., ozone) in regional airsheds to develop plans (SIPs) and rules with the objective of attaining ambient air quality standards (NAAQS). Both atmospheric dynamics and chemistry are important components of 3-D grid models. In contrast to mechanistic plume and grid models, receptor models are statistical; they determine the statistical contribution of various sources to pollutant concentrations at a given location based on the relative amounts of pollutants at source and receptor. Most air quality models are deterministic.

Air Quality Models

Model	Type	Use	Additional Information
CMAQ	3-D Grid	SIP development, NAAQS setting (CAA). The model provides estimates of ozone, particulates, toxics, and acid deposition and simulates chemical and physical properties related to atmospheric trace gas transformations and distributions. Model has three components including, meteorological system, an emissions model for estimating anthropogenic and natural emissions, and a chemistry-transport modeling system.	EPA 2007d Byun and Ching 1999
UAM	3-D Grid	Model calculates concentrations of inert and chemically reactive pollutants and is used to evaluate air quality, particularly related to ambient ozone concentrations.	Systems Applications International, Inc., 1999

(Continued)

Air Quality Models Continued

Model	Type	Use	Additional Information
REMSAD	3-D Grid	Using simulation of physical and chemical processes in the atmosphere that impact pollutant concentrations, model calculates concentration of inert and chemically reactive pollutants.	ICF International/ Systems Applications International 2006, ICF Consulting 2005
ICSC	Plume	PSD permitting; toxics exposure (CAA, TSCA).	
CALPUFF		Non-steady-state air quality dispersion model that simulates long range transport of pollutants.	
CMB	Receptor	Relative contributions of sources. Receptor model used for air resource management purposes.	EPA 2006f Coulter 2004

EXPOSURE MODELS

The primary objective of exposure models is to estimate the dose of pollutant which humans or animals are exposed to via inhalation, ingestion and/or dermal uptake. These models bridge the gap between concentrations of pollutants in the environment and the doses humans receive based on their activity. Pharmacokinetic models take this one step further and estimate dose to tissues in the body. Since exposure is inherently tied to behavior, exposure models may also simulate activity, for example a model that estimates dietary consumption of pollutants. In addition to the Lifeline model described below, other examples of models that estimate dietary exposure to pesticides include Calendex and CARES. These models can be either deterministic or probabilistic, but are well-suited for probabilistic methods due to the variability of activity within a population.

Exposure Models

Model	Type	Use	Additional Information
Lifeline	Diet, water and dermal of single chemical	Aggregate dose of pesticide via multiple pathways.	Lifeline Group, Inc. 2007 Lifeline Group, Inc. 2006

IEUBK	Multipath-way, single chemical	Dose of lead to children's blood via multiple pathways. Estimates exposure from lead in media (air, water, soil, dust, diet, and paint and other sources) using pharmacokinetic models to predict blood lead levels in children 6 months to 7 years old. The model can be used as a tool for the determination of site-specific cleanup levels.	EPA 2005c EPA 1994
Air Pollutants Exposure Model (APEX)	Inhalation exposure model	Simulates an individual's exposure to an air pollutant and their movement through space and time in indoor or outdoor environments. Provides dose estimates and summary exposure information for each individual.	EPA 2007e Richmond et al. 2001

HUMAN HEALTH AND ENVIRONMENT RESPONSE MODELS

Human Health Effects Models

Health effects models provide a statistical relationship between a dose of a chemical and an adverse human health effect. Health effects models are statistical methods, hence models in this category are almost exclusively empirical. They can be further classified as toxicological and epidemiological. The former refer to models derived from observations in controlled experiments, usually with nonhuman subjects. The latter refer to models derived from observations over large populations. Health models use statistical methods and assumptions that ultimately assume cause and effect. Included in this category are models that extrapolate information from non-human subject experiments. Also, physiologically based pharmacokinetic models can help predict human toxicity to contaminants through mathematical modeling of absorption, distribution, storage, metabolism, and excretion of toxicants.

The output from health models is almost always a dose, such as a safe level (for example, reference dose [RfD]), a cancer potency index (CPI), or an expected health end point (for example, lethal dose for 50% of the population (LD_{50}) or number of asthma cases). There also exist model *applications* that facilitate the use of the statistical methods.

Human Health Effects Models

Model	Type	Use	Additional Information
Benchmark dose model	Software tool for applying a variety of statistical models to analyze dose-response data	To estimate risk of pollutant exposure. Models fit to dose-response data to determine a benchmark dose that is associated with a particular benchmark response.	EPA 2007f EPA 2000
Linear Cancer model	Statistical analysis method	To estimate the risk posed by carcinogenic pollutants.	

Ecological Effects Models

Ecological effects models, like human health effects models, define relationships between a level of pollutant exposure and a particular ecological indicator. Many ecological effects models simulate aquatic environments, and ecological indicators are related directly to environmental concentrations. Examples of ecological effects indicators that have been modeled are: algae blooms, BOD, fish populations, crop yields, coast line erosion, lake acidity, and soil salinity.

Ecological Effects Models

Model	Type	Use	Additional Information
AQUATOX	Integrated fate and effects of pollutants in aquatic environment	Ecosystem model that predicts the environmental fate of chemicals in aquatic ecosystems, as well as direct and indirect effects on the resident organisms. Potential applications to management decisions include water quality criteria and standards, TMDLs, and ecological risk assessments of aquatic systems.	EPA 2006g Hawkins 2005, Rashleigh 2007
BASS	Simulates fish populations exposed to pollutants (mechanistic)	Models dynamic chemical bioconcentration of organic pollutants and metals in fish. Estimates are being used for ecological risks to fish in addition to realistic dietary exposures to humans and wildlife.	EPA 2006h

SERAFM	Steady-state modeling system used to predict mercury concentration in wildlife	Predicts total mercury concentrations in fish and speciated mercury concentrations in water and sediments.	EPA 2007g Knightes 2005
PATCH	Movement of invertebrates in their habitat	Provides population estimates of territorial terrestrial vertebrate species over time, in addition to survival and fecundity rates, and orientation of breeding sites. Determine ecological effects of regulation.	EPA 2007h Lawler et al. 2006

ECONOMIC IMPACT MODELS

This category includes a broad group of models that are used in many different aspects of EPA's activities including: rulemaking (regulatory impact assessments), priority setting, enforcement, and retrospective analyses. Models that produce a dollar value as output belong in this category. Models can be divided into cost models, which may include or exclude behavior responses, and benefit models. The former incorporate economic theory on how markets (supply, demand, and pricing) will respond as a result of an action.

Economic models are traditionally deterministic, though there is a trend toward greater use of uncertainty methods in cost-benefit analysis.

Economic Impact Models

Model	Type	Use	Additional Information
ABEL	Micro Economic	Assess a single firm's ability to pay compliance costs or fees. Estimates claims from defendants that they cannot afford to pay for compliance, clean-up or civil penalties using information from tax return data and cash-flow analysis. Used for settlement negotiations.	EPA 1999
Nonroad Diesel Economic	Macro economic for impact	Multimarket model to analyze how producers and consumers are expected to respond to compliance costs associated with the rule.	EPA 2005d

(Continued)

Economic Impact Models Continued

Model	Type	Use	Additional Information
Impact Model (NDEIM)	of the nonroad diesel emissions standards rule	Estimates and stratifies emissions for nonroad equipment. Model can be used to inform State Implementation Plans and regulatory analyses.	
BenMAP	Noneco-nomic and economic benefits from air quality	Model that estimates the health benefits associated with air quality changes by estimating changes in incidences of a wide range of health outcomes and then placing an economic value on these reduced incidences.	EPA 2007i

NONECONOMIC IMPACT MODELS

Noneconomic impact models evaluate the effects of contaminants on a variety of noneconomic parameters, such as on crop yields and buildings. Note that other noneconomic impacts, such as impacts on human health or ecosystems, are derived from the human health and ecological effects models discussed previously.

Noneconomic Impact Models

Model	Type	Use	Additional Information
TDM (Travel Demand Management)	Model used to evaluate travel demand management strategies	Evaluates travel demand management strategies to determine vehicle-trip reduction effects. Model used to support transit policies including HOV lanes, carpooling, telecommuting, and pricing and travel subsidies.	Shiftan and Suhrbier 2002
CERES-Wheat	Crop-growth yield model	Simulates effects of planting density, weather, water, soil, and nitrogen on crop growth, development, and yield. Predicts management strategies that impact crop yield.	Ritchie and Godwin 2007
PHREEQE-A	Models effects of acidification on stone	Simulates the effects of acidic solutions on carbonate stone.	Parkhurst et al. 1980

REFERENCES

Birgand, F. 2004. Evaluation of QUAL2E. Pp. 99-106 in Agricultural Non-Point Source Water Quality Models: Their Use and Application, J.E. Parsons, D.L. Thomas, and R.L. Huffman, eds. Southern Cooperative Series Bulletin 398 [online]. Available: http://www3.bae.ncsu.edu/Regional-Bulletins/Modeling-Bulletin/qual2e.html.

Brown, L.C. 1986. Uncertainty Analysis Using QUAL2E. EPA/600/D/86/053. Office of Research and Development, U.S. Environmental Protection Agency.

Brown, L.C., and T. O. Barnwell. 1987. The Enhanced Stream Water Quality Models QUAL2E and QUAL2E-UNCAS: Documentation and User Manual. EPA/600/3-87/007. Environmental Research Laboratory, U.S. Environmental Protection Agency, Athens, GA.

Byun, D.W., and J.K.S. Ching. 1999. Science Algorithms of the EPA Models-3 Community Multiscale Air Quality (CMAQ) Modeling System. EPA/600/R-99/030. Atmospheric Modeling Division, National Exposure Research Laboratory, U.S. Environmental Protection Agency, Research Triangle Park, NC [online]. Available: http://www.epa.gov/asmdnerl/CMAQ/CMAQscienceDoc.html [accessed June 13, 2007].

Caliper Corporation. 2007. TransCAD [online]. Available: http://www.caliper.com/tcovu.htm [accessed June 13, 2007].

Coulter, C.T. 2004. EPA-CMB8.2 Users Manual. EPA-452/R-04-011. Air Quality Modeling Group, Emissions, Monitoring & Analysis Division, Office of Air Quality Planning & Standards, Research Triangle Park, NC [online]. Available: http://www.epa.gov/scram001/models/receptor/EPA CMB82Manual.pdf [accessed June 13, 2007].

Donigian, A.S., Jr. 2002. Watershed Model Calibration and Validation: The HSPF Experience. WEF National TMDL Science and Policy 2002, November 13-16, 2002. Phoenix, AZ [online]. Available: http://hspf.com/TMDL.Nov02.Donigian.Paper.doc [accessed June 13, 2007].

EIA (Energy Information Administration). 1993. Documentation of the DRI Model of the U.S. Economy. [online]. Available: tonto.eia.doe.gov/FTP ROOT/modeldoc/m061.pdf [accessed March 31, 2007].

EPA (U.S. Environmental Protection Agency). 1994. Guidance Manual for the Integrated Exposure Uptake Biokinetic Model for Lead in Children. EPA/540/R-93/081. OSWER9285.7-15-1. PB93-963510. Office of Solid Waste and Emergency Response, U.S. Environmental Protection Agency, Washington, DC. February 1994 [online]. Available: http://www.epa.gov/superfund/programs/lead/products.htm [accessed Nov. 2, 2006].

EPA (U.S. Environmental Protection Agency). 1998. BIOPLUME III: Natural Attenuation Decision Support System. User's Manual Version 1.0. EPA/600/R-98/010. U.S. Environmental Protection Agency, Washington, DC.

EPA (U.S. Environmental Protection Agency). 1999. ABEL Model. Environmental Data Registry, U.S. Environmental Protection Agency [online]. Available: http://iaspub.epa.gov/edr/edr_proc_qry.navigate?P_LIST_OPTION_CD=CSDIS&P_REG_AUTH_IDENTIFIER=1&P_DATA_IDENTIFIER=90389&P_VERSION=1 [accessed June 14, 2007].
EPA (U.S. Environmental Protection Agency). 2000. Benchmark Dose Technical Guidance Document. EPA/630/R-00/001. External Review Draft. Risk Assessment Forum, U.S. Environmental Protection Agency, Washington, DC [online]. Available: http://www.epa.gov/nceawww1/pdfs/bmds/BMD External_10_13_2000. pdf [accessed June 12, 2007].
EPA (U.S. Environmental Protection Agency). 2001. PLOAD Version 3.0: An ArcView GIS Tool to Calculate Nonpoint Sources of Pollution in Watershed and Stormwater Projects. User's Manual. Office of Water Science, U.S. Environmental Protection Agency. January 2001 [online]. Available: www.epa.gov/waterscience/basins/b3docs/PLOAD_v3.pdf [accessed March 21, 2007].
EPA (U.S. Environmental Protection Agency). 2004. MOVES2004 User Guide. Draft. EPA420-P-04-019. Assessment and Standards Division, Office of Transportation and Air Quality, U.S. Environmental Protection Agency, Washington, DC. November 2004 [online]. Available: http://www.epa.gov/otaq/models/ngm/420p04019.pdf [accessed June 13, 2007].
EPA (U.S. Environmental Protection Agency). 2005a. User's Guide for the Final NONROAD2005 Model. EPA420-R-05-013. Assessment and Standards Division, Office of Transportation and Air Quality, U.S. Environmental Protection Agency, Washington, DC. December 2005 [online]. Available: http://www.epa.gov/otaq/models/nonrdmdl/nonrdmdl2005/420r05013.pdf [accessed June 13, 2007].
EPA (U.S. Environmental Protection Agency). 2005b. PRZM-3, A Model for Predicting Pesticide and Nitrogen Fate in the Crop Root and Unsaturated Soil Zones: Users Manual for Release 3.12.2. EPA/600/R-05/111. U.S. Environmental Protection Agency, Washington, DC.
EPA (U.S. Environmental Protection Agency). 2005c. Integrated Exposure Uptake Biokinetic Model for Lead in Children, Windows Version (IEUBKwin v1.0 build 263). Superfund, U.S. Environmental Protection Agency. December 2005 [online]. Available: http://www.epa.gov/superfund/programs/lead/products.htm#ieubk [accessed June 13, 2007].
EPA (U.S. Environmental Protection Agency). 2005d. Economic Impact Analysis of the Standards of Performance for Stationary Compression Ignition Internal Combustion Engines. EPA-452/R-05-006. Office of Air Quality Planning and Standards, U.S. Environmental Protection Agency, Research Triangle Park, NC. June 2005 [online]. Available: http://www.epa.gov/ttn/atw/nsps/cinsps/ci_nsps_eia_reportfinalforproposal.pdf [accessed June 14, 2007].
EPA (U.S. Environmental Protection Agency). 2006a. Biogenic Emissions Inventory System (BEIS) Modeling. Atmospheric Sciences Modeling Divi-

sion, Office of Research and Development, U.S. Environmental Protection Agency [online]. Available: http://www.epa.gov/asmdnerl/biogen.html [accessed June 13, 2007].

EPA (U.S. Environmental Protection Agency). 2006b. NONROAD Model (Nonroad Engines, Equipment, and Vehicles). Office of Transportation and Air Quality, U.S. Environmental Protection Agency [online]. Available: http://www.epa.gov/otaq/nonrdmdl.htm [accessed June 13, 2007].

EPA (U.S. Environmental Protection Agency). 2006c. CSMoS Ground-Water Modeling Software. Ground Water Technical Support Center, Risk Management Research, Office of Research and Development, U.S. Environmental Protection Agency [online]. Available: http://www.epa.gov/ada/csmos/models.html [accessed June 13, 2007].

EPA (U.S. Environmental Protection Agency). 2006d. HSPF. Exposure Assessment Models. U.S. Environmental Protection Agency [online]. Available: http://www.epa.gov/ceampubl/swater/hspf/ [accessed June 13, 2007].

EPA (U.S. Environmental Protection Agency). 2006e. Water Quality Analysis Simulation Program (WASP). Ecosystems Research Division, Office of Research and Development, U.S. Environmental Protection Agency [online]. Available: http://www.epa.gov/athens/wwqtsc/html/wasp.html [accessed June 13, 2007].

EPA (U.S. Environmental Protection Agency). 2006f. The Chemical Mass Balance (CMB) Model EPA-CMBv8.2. Receptor Miodeling, Air Quality Models, U.S. Environmental Protection Agency [online]. Available: http://www.epa.gov/scram001/receptor_cmb.htm [accessed June 13, 2007].

EPA (U.S. Environmental Protection Agency). 2006g. AQUATOX. Office of Water Science, U.S. Environmental Protection Agency [online]. Available: http://www.epa.gov/waterscience/models/aquatox/ [accessed June 14, 2007].

EPA (U.S. Environmental Protection Agency). 2006h. BASS. Ecosystems Research Division, Office of Research and Development, U.S. Environmental Protection Agency [online]. Available: http://www.epa.gov/athens/research/modeling/bass.html [accessed June 14, 2007].

EPA (U.S. Environmental Protection Agency). 2007a. Better Assessment Science Integrating Point & Nonpoint Sources (BASINS). Water Quality Models and Tools, Office of Water Science, U.S. Environmental Protection Agency [online]. Available: http://www.epa.gov/waterscience/basins/ [accessed June 13, 2007].

EPA (U.S. Environmental Protection Agency). 2007b. MOBILE6 Vehicle Emission Modeling Software. Office of Transportation and Air Quality, U.S. Environmental Protection Agency [online]. Available: http://www.epa.gov/otaq/m6.htm [accessed June 13, 2007].

EPA (U.S. Environmental Protection Agency). 2007c. Modeling Products. Exposure Assessment, U.S. Environmental Protection Agency [online].

Available: http://www.epa.gov/ceampubl/products.htm [accessed June 13, 2007].

EPA (U.S. Environmental Protection Agency). 2007d. Community Multiscale Air Quality (CMAQ). Atmospheric Science Modeling, Office of Research and Development, U.S. Environmental Protection Agency [online]. Available: http://www.epa.gov/asmdnerl/CMAQ/index.html [accessed June 13, 2007].

EPA (U.S. Environmental Protection Agency). 2007e. Human Exposure Modeling—Air Pollutants Exposure Model (APEX/TRIM.$Expo_{Inhalation}$). Office of Air and Radiation, U.S. Environmental Protection Agency [online]. Available: http://www.epa.gov/ttn/fera/human_apex.html [accessed June 13, 2007].

EPA (U.S. Environmental Protection Agency). 2007f. Benchmark Dose Software. National Center for Environmental Assessment, Office of Research and Development, U.S. Environmental Protection Agency [online]. Available: http://cfpub.epa.gov/ncea/cfm/recordisplay.cfm?deid=164443 [accessed June 14, 2007].

EPA (U.S. Environmental Protection Agency). 2007g. SERAFM—Spreadsheet-based Ecological Risk Assessment for the Fate of Mercury. Surface water Models. Exposure Assessment Models, U.S. Environmental Protection Agency [online]. Available: http://www.epa.gov/ceampubl/swater/serafm/index.htm [accessed June 14, 2007].

EPA (U.S. Environmental Protection Agency). 2007h. Program to Assist in Tracking Critical Habitat (PATCH). Western Ecology Division, Office of Research and Development, U.S. Environmental Protection Agency, Corvallis, OR [online]. Available: http://www.epa.gov/wed/pages/models/patch/patchmain.htm [accessed June 14, 2007].

EPA (U.S. Environmental Protection Agency). 2007i. Benefits Analysis Models/Tools. Economics & Cost Analysis Support, U.S. Environmental Protection Agency [online]. Available: http://www.epa.gov/ttnecas1/benmodels.html [accessed June 14, 2007].

Glover, E.L., and M. Cumberworth. 2003. MOBILE6.1 Particulate Emission Factor Model Technical Description—Final Report. EPA420-R-03-001. Assessment and Standards Division, Office of Transportation and Air Quality, U.S. Environmental Protection Agency. January 2003 [online]. Available: http://www.epa.gov/otaq/models/mobile6/r03001.pdf [accessed June 13, 2007].

Hawkins, T. 2005. Critical Evaluation of the Aquatox Model. Carnegie Mellon University [online]. Available: http://www.ce.cmu.edu/~trh/Professional/Research/Hawkins_CriticalEvaluationOfTheAquatoxModel.pdf [accessed March 31, 2007].

Hayes, J.T., P.A. O'Rourke, W.H. Terjung, and P.E. Todhunter. 1982. A feasible crop yield model for worldwide international food production. Int. J. Biometeorol. 26(3):239-257.

ICF Consulting. 2005. User's Guide to the Regional Modeling System for Aerosols and Deposition (REMSAD): Version 8 [online.] Available: http://www.remsad.com/documents/remsad_users_guide_v8.00_112305.pdf [accessed March 31, 2007].

ICF International/Systems Applications International. 2006. Regional Modeling System for Aerosols and Deposition (REMSAD) [online]. Available: http://www.remsad.com/ [accessed June 13, 2007].

Knightes, C.D. 2005. SERAFM: An Ecological Risk Assessment Tool for Evaluating Wildlife Exposure Risk Associated with Mercury-Contaminated Sediment in Lake and River Systems. Presentation at EPA Science Forum 2005, May 16-18, 2005, Washington, DC.

Lawler, J.J., D. White, R.P. Neilson, and A.R. Blaustein. 2006. Predicting climate-induced range shifts: Model differences and model reliability. Glob. Change Biol. 12(8):1568-1584.

Lifeline Group, Inc. 2006. Users Manual: LifeLine™ Version 4.3. Software for Modeling Aggregate and Cumulative Exposures to Pesticides and Chemicals. April 5, 2006 [online]. Available: http://www.thelifelinegroup.org/lifeline/documents/v4.3_UserManual.pdf [accessed March 31, 2007].

Lifeline Group, Inc. 2007. Lifeline Software [online]. Available: http://www.thelifelinegroup.org/index.htm [accessed June 13, 2007].

MBL (Marine Biological Laboratory). 2005. Marine Biological Laboratory General Ecosystem Model (MBL-GEM). The Ecosystem Center, Marine Biological Laboratory, MA [online]. Available: http://ecosystems.mbl.edu/Research/Models/gem/welcome.html [accessed June 13, 2007].

MIT (Massachusetts Institute of Technology). 2006. Natural Emissions Model (NEM). The MIT Integrated Global System Model: Ecosystems Impacts [online]. Available: http://web.mit.edu/globalchange/www/tem.html#nem [accessed June 13, 2007].

Parkhurst, D.L., D.C. Thoratenson, L.N. Plummer. 1980. PHREEQE: A Computer Program for Geochemical Calculations. U.S. Geological Survey Water Research Investigations 80-96. Reston, VA: U.S. Geological Survey.

Prudic, D.E., L.F. Konikow, and E.R. Banta. 2004. A New Stream- Flow Routing (SFR1) Package to Simulate Stream-Aquifer Interaction with MODFLOW-2000. U.S. Geological Survey Open-File Report 2004-1042. U.S. Department of the Interior, U.S. Geological Survey [online]. Available: http://pubs.usgs.gov/of/2004/1042/ [accessed June 13, 2007].

Rashleigh, B. 2007. Assessment of lake ecosystem response to toxic events with the AQUATOX model. Pp. 293-299 in Assessment of the Fates and Effects of Toxic Agents on Water Resources, I.E. Gonenc, V. Koutitonsky, B. Rashleigh, R.A. Ambrose, and J.P. Wolfin, eds. Dordrecht, The Netherlands: Springer.

Richmond, H., T. Palma, G. Glen, and L. Smith. 2001. Overview of APEX (2.0): EPA's Pollutant Exposure Model for Criteria and Air Toxic Inhalation Exposures. Annual Meeting of the International Society of Exposure Analysis, November 4-8, 2001. Charleston, SC.

Ritchie, J.T., and D. Godwin. 2007. CARES Wheat 2.0. Michigan State University [online]. Available: http://nowlin.css.msu.edu/wheat_book/ [accessed June 14, 2007].

Schwarz, G.E., A.B. Hoos, R.B. Alexander, and R.A. Smith. 2006. The SPARROW Surface Water-Quality Model: Theory, Application and User Documentation. U.S. Geological Survey Techniques and Methods 6-B3. U.S. Geological Survey [online]. Available: http://pubs.usgs.gov/tm/2006/tm6b3/PDF.htm. [accessed March 31, 2007].

Shiftan, Y., and J. Suhrbier. 2002. The analysis of travel and emission impacts of travel demand management strategies using activity-based models. Transportation 29(2):145-168.

Sokolov, A.P., C.A. Schlosser, S. Dutkiewicz, S. Paltsev, D.W. Kicklighter, H.D. Jacoby, R.G. Prinn, C.E. Forest, J. Reilly, C. Wang, B. Felzer, M.C. Sarofim, J. Scott, P.H. Stone, J.M. Melillo, and J. Cohen. 2005. The MIT Integrated Global System Model (IGSM) Version 2: Model Description and Baseline Evaluation. Report No. 124. Joint Program on the Science and Policy of Global Change, Massachusetts Institute of Technology [online]. Available: http://web.mit.edu/globalchange/www/abstracts.html#a124 [accessed March 31, 2007].

Systems Applications International, Inc. 1999. User's Guide to the Variable-Grid Urban Airshed Model (UAM-V). Systems Applications International, Inc., San Rafael, CA [online]. Available: http://www.uamv.com/documents/uam-v_1.31_user's_guide.pdf [accessed June 13, 2007].

USGS (U.S. Geological Survey). 2007a. SPARROW Modeling of Surface-Water Quality. U.S. Department of the Interior, U.S. Geological Survey [online]. Available: http://water.usgs.gov/nawqa/sparrow/ [accessed June 13, 2007].

USGS (U.S. Geological Survey). 2007b. MODFLOW-2000 Version 1.17.02. USGS Ground-Water Software. U.S. Department of the Interior, U.S. Geological Survey [online]. Available: http://water.usgs.gov/nrp/gwsoftware/modflow2000/modflow2000.html [accessed June 13, 2007].

Vukovich, J.M., and T. Pierce. 2002. The Implementation of BEIS3 within the SMOKE. 11th International Emission Inventory Conference: Emission Inventories—Partnering for the Future, April 15-18, 2002, Atlanta, GA [online]. Available: http://www.epa.gov/ttn/chief/conference/ei11/modeling/vukovich.pdf [accessed June 13, 2007].

Wilson, J.D., and R.L. Naff. 2004. MODFLOW-2000: The U.S. Geological Survey Modular Ground-Water Model-GMG Linear Equation Solver Package Documentation. U.S. Geological Survey Water Resources Open-File Report 2004-1261. U.S. Department of the Interior, U.S. Geological Survey [online]. Available: http://pubs.usgs.gov/of/2004/1261/ [accessed June 13, 2007].

Young, T., R. Randolph, and D. Bowman. 1994. Economic Growth Analysis System: Version 2.0. EPA/600/SR-94/139. Air and Energy Engineering, Research Laboratory, U.S. Environmental Protection Agency, Research

Triangle Park [online]. Available: http://www.p2pays.org/ref/07/0622.pdf [accessed June 13, 2007].